AnswerMan HVAC&R Reference Guide

Printed in

PREFACE

The Esco Institute has made a serious effort to provide accurate information in Answer Man. However, the probability exists that there are errors and misprints and that variations in data values may also occur depending on field conditions.

Information included in this book should only be considered as a general guide and the Esco Institute does not represent the information as being exact.

The Esco Institute would appreciate being notified of any errors, omissions, or misprints which may occur in Answer Man. Your suggestions for future editions would also be greatly appreciated.

The information contained in Answer Man was collected from many different sources and numerous hours of research and hard work have gone into selecting the most valuable data to be included in this edition. We look forward to your comments as to future editions.

Our special thanks to Gene Anderson, Jim Edge, Dianne Guy, Walter Podrazik and Jerry Weiss for their contributions in the Answer Man publication.

PHONE NUMBERS

2006

	S M T W T F S	S M T W T F S	S M T W T F S	S M T W T F S

JANUARY

S M T W T F S
1 2 3 4 5 6 7
8 9 10 11 12 13 14
15 16 17 18 19 20 21
22 23 24 25 26 27 28
29 30 31

FEBRUARY

| 1 2 3 4 |
| 5 6 7 8 9 10 11 |
| 12 13 14 15 16 17 18 |
| 19 20 21 22 23 24 25 |
| 26 27 28 |

MARCH

| 1 2 3 4 |
| 5 6 7 8 9 10 11 |
| 12 13 14 15 16 17 18 |
| 19 20 21 22 23 24 25 |
| 26 27 28 29 30 31 |

APRIL

| 1 |
| 2 3 4 5 6 7 8 |
| 9 10 11 12 13 14 15 |
| 16 17 18 19 20 21 22 |
| 23 24 25 26 27 28 29 |
| 30 |

MAY

| 1 2 3 4 5 6 |
| 7 8 9 10 11 12 13 |
| 14 15 16 17 18 19 20 |
| 21 22 23 24 25 26 27 |
| 28 29 30 31 |

JUNE

| 1 2 3 |
| 4 5 6 7 8 9 10 |
| 11 12 13 14 15 16 17 |
| 18 19 20 21 22 23 24 |
| 25 26 27 28 29 30 |

JULY

| 1 |
| 2 3 4 5 6 7 8 |
| 9 10 11 12 13 14 15 |
| 16 17 18 19 20 21 22 |
| 23 24 25 26 27 28 29 |
| 30 31 |

AUGUST

| 1 2 3 4 5 |
| 6 7 8 9 10 11 12 |
| 13 14 15 16 17 18 19 |
| 20 21 22 23 24 25 26 |
| 27 28 29 30 31 |

SEPTEMBER

| 1 2 |
| 3 4 5 6 7 8 9 |
| 10 11 12 13 14 15 16 |
| 17 18 19 20 21 22 23 |
| 24 25 26 27 28 29 30 |

OCTOBER

| 1 2 3 4 5 6 7 |
| 8 9 10 11 12 13 14 |
| 15 16 17 18 19 20 21 |
| 22 23 24 25 26 27 28 |
| 29 30 31 |

NOVEMBER

| 1 2 3 4 |
| 5 6 7 8 9 10 11 |
| 12 13 14 15 16 17 18 |
| 19 20 21 22 23 24 25 |
| 26 27 28 29 30 |

DECEMBER

| 1 2 |
| 3 4 5 6 7 8 9 |
| 10 11 12 13 14 15 16 |
| 17 18 19 20 21 22 23 |
| 24 25 26 27 28 29 30 |
| 31 |

2007

JANUARY

| 1 2 3 4 5 6 |
| 7 8 9 10 11 12 13 |
| 14 15 16 17 18 19 20 |
| 21 22 23 24 25 26 27 |
| 28 29 30 31 |

FEBRUARY

| 1 2 3 |
| 4 5 6 7 8 9 10 |
| 11 12 13 14 15 16 17 |
| 18 19 20 21 22 23 24 |
| 25 26 27 28 |

MARCH

| 1 2 3 |
| 4 5 6 7 8 9 10 |
| 11 12 13 14 15 16 17 |
| 18 19 20 21 22 23 24 |
| 25 26 27 28 29 30 31 |

APRIL

| 1 2 3 4 5 6 7 |
| 8 9 10 11 12 13 14 |
| 15 16 17 18 19 20 21 |
| 22 23 24 25 26 27 28 |
| 29 30 |

MAY

| 1 2 3 4 5 |
| 6 7 8 9 10 11 12 |
| 13 14 15 16 17 18 19 |
| 20 21 22 23 24 25 26 |
| 27 28 29 30 31 |

JUNE

| 1 2 |
| 3 4 5 6 7 8 9 |
| 10 11 12 13 14 15 16 |
| 17 18 19 20 21 22 23 |
| 24 25 26 27 28 29 30 |

JULY

| 1 2 3 4 5 6 7 |
| 8 9 10 11 12 13 14 |
| 15 16 17 18 19 20 21 |
| 22 23 24 25 26 27 28 |
| 29 30 31 |

AUGUST

| 1 2 3 4 |
| 5 6 7 8 9 10 11 |
| 12 13 14 15 16 17 18 |
| 19 20 21 22 23 24 25 |
| 26 27 28 29 30 31 |

SEPTEMBER

| 1 |
| 2 3 4 5 6 7 8 |
| 9 10 11 12 13 14 15 |
| 16 17 18 19 20 21 22 |
| 23 24 25 26 27 28 29 |
| 30 |

OCTOBER

| 1 2 3 4 5 6 |
| 7 8 9 10 11 12 13 |
| 14 15 16 17 18 19 20 |
| 21 22 23 24 25 26 27 |
| 28 29 30 31 |

NOVEMBER

| 1 2 3 |
| 4 5 6 7 8 9 10 |
| 11 12 13 14 15 16 17 |
| 18 19 20 21 22 23 24 |
| 25 26 27 28 29 30 |

DECEMBER

| 1 |
| 2 3 4 5 6 7 8 |
| 9 10 11 12 13 14 15 |
| 16 17 18 19 20 21 22 |
| 23 24 25 26 27 28 29 |
| 30 31 |

2008

	JANUARY		FEBRUARY		MARCH		APRIL

JANUARY

S	M	T	W	T	F	S
		1	2	3	4	5
6	7	8	9	10	11	12
13	14	15	16	17	18	19
20	21	22	23	24	25	26
27	28	29	30	31		

FEBRUARY

S	M	T	W	T	F	S
					1	2
3	4	5	6	7	8	9
10	11	12	13	14	15	16
17	18	19	20	21	22	23
24	25	26	27	28	29	

MARCH

S	M	T	W	T	F	S
						1
2	3	4	5	6	7	8
9	10	11	12	13	14	15
16	17	18	19	20	21	22
23	24	25	26	27	28	29
30	31					

APRIL

S	M	T	W	T	F	S
		1	2	3	4	5
6	7	8	9	10	11	12
13	14	15	16	17	18	19
20	21	22	23	24	25	26
27	28	29	30			

MAY

S	M	T	W	T	F	S
				1	2	3
4	5	6	7	8	9	10
11	12	13	14	15	16	17
18	19	20	21	22	23	24
25	26	27	28	29	30	31

JUNE

S	M	T	W	T	F	S
1	2	3	4	5	6	7
8	9	10	11	12	13	14
15	16	17	18	19	20	21
22	23	24	25	26	27	28
29	30					

JULY

S	M	T	W	T	F	S
		1	2	3	4	5
6	7	8	9	10	11	12
13	14	15	16	17	18	19
20	21	22	23	24	25	26
27	28	29	30	31		

AUGUST

S	M	T	W	T	F	S
					1	2
3	4	5	6	7	8	9
10	11	12	13	14	15	16
17	18	19	20	21	22	23
24	25	26	27	28	29	30
31						

SEPTEMBER

S	M	T	W	T	F	S
	1	2	3	4	5	6
7	8	9	10	11	12	13
14	15	16	17	18	19	20
21	22	23	24	25	26	27
28	29	30				

OCTOBER

S	M	T	W	T	F	S
			1	2	3	4
5	6	7	8	9	10	11
12	13	14	15	16	17	18
19	20	21	22	23	24	25
26	27	28	29	30	31	

NOVEMBER

S	M	T	W	T	F	S
						1
2	3	4	5	6	7	8
9	10	11	12	13	14	15
16	17	18	19	20	21	22
23	24	25	26	27	28	29
30						

DECEMBER

S	M	T	W	T	F	S
	1	2	3	4	5	6
7	8	9	10	11	12	13
14	15	16	17	18	19	20
21	22	23	24	25	26	27
28	29	30	31			

2009

JANUARY

S	M	T	W	T	F	S
				1	2	3
4	5	6	7	8	9	10
11	12	13	14	15	16	17
18	19	20	21	22	23	24
25	26	27	28	29	30	31

FEBRUARY

S	M	T	W	T	F	S
1	2	3	4	5	6	7
8	9	10	11	12	13	14
15	16	17	18	19	20	21
22	23	24	25	26	27	28

MARCH

S	M	T	W	T	F	S
1	2	3	4	5	6	7
8	9	10	11	12	13	14
15	16	17	18	19	20	21
22	23	24	25	26	27	28
29	30	31				

APRIL

S	M	T	W	T	F	S
			1	2	3	4
5	6	7	8	9	10	11
12	13	14	15	16	17	18
19	20	21	22	23	24	25
26	27	28	29	30		

MAY

S	M	T	W	T	F	S
					1	2
3	4	5	6	7	8	9
10	11	12	13	14	15	16
17	18	19	20	21	22	23
24	25	26	27	28	29	30
31						

JUNE

S	M	T	W	T	F	S
	1	2	3	4	5	6
7	8	9	10	11	12	13
14	15	16	17	18	19	20
21	22	23	24	25	26	27
28	29	30				

JULY

S	M	T	W	T	F	S
			1	2	3	4
5	6	7	8	9	10	11
12	13	14	15	16	17	18
19	20	21	22	23	24	25
26	27	28	29	30	31	

AUGUST

S	M	T	W	T	F	S
						1
2	3	4	5	6	7	8
9	10	11	12	13	14	15
16	17	18	19	20	21	22
23	24	25	26	27	28	29
30	31					

SEPTEMBER

S	M	T	W	T	F	S
		1	2	3	4	5
6	7	8	9	10	11	12
13	14	15	16	17	18	19
20	21	22	23	24	25	26
27	28	29	30			

OCTOBER

S	M	T	W	T	F	S
				1	2	3
4	5	6	7	8	9	10
11	12	13	14	15	16	17
18	19	20	21	22	23	24
25	26	27	28	29	30	31

NOVEMBER

S	M	T	W	T	F	S
1	2	3	4	5	6	7
8	9	10	11	12	13	14
15	16	17	18	19	20	21
22	23	24	25	26	27	28
29	30					

DECEMBER

S	M	T	W	T	F	S
		1	2	3	4	5
6	7	8	9	10	11	12
13	14	15	16	17	18	19
20	21	22	23	24	25	26
27	28	29	30	31		

TABLE OF CONTENTS

6. GENERAL SCIENCE

7. CONVERSION TABLES AND EQUIVALENTS

REFRIGERANTS
and
PROPERTIES

SATURATION CHART

Italics Indicate inches Mercury   Pressure numbers indicate psig

TEMP.		REFRIGERANT NO. & CYL. COLOR CODE				
F°	C°	Silver 717	Orange 11	White 12	Green 22	Purple 113
-50	-45.6	*14.3*	*28.9*	*15.4*	*6.2*	
-45	-42.8	*11.7*	*28.7*	*13.3*	*3.0*	
-40	-40.0	*8.7*	*28.4*	*11.0*	*0.5*	
-35	-37.2	*5.4*	*28.1*	*8.4*	2.5	
-30	-34.4	*1.6*	*27.8*	*5.5*	4.8	*29.3*
-25	-31.7	1.3	*27.4*	*2.3*	7.3	*29.2*
-20	-28.9	3.6	*27.0*	*0.6*	10.1	*29.1*
-18	-27.8	4.6	*26.8*	1.3	11.3	*29.0*
-16	-26.7	5.6	*26.6*	2.1	12.5	*29.0*
-14	-25.6	6.7	*26.4*	2.8	13.8	*28.9*
-12	-24.4	7.9	*26.2*	3.7	15.1	*28.8*
-10	-23.3	9.0	*26.0*	4.5	16.5	*28.7*
-08	-22.2	10.3	*25.8*	5.4	17.9	*28.6*

F°	C°	R123	Blue R134a	Yellow 500	Orchid 502	Aqua 503
-06	-21.1	*26.5*	3.7	9.9	25.8	207.3
-04	-20.0	*26.3*	4.6	11.0	27.5	214.7
-02	-18.9	*26.1*	5.5	12.1	29.3	222.3
0	-17.8	*25.8*	6.5	13.3	31.1	230.0
2	-16.7	*25.6*	7.5	14.5	33.0	238.0
4	-15.6	*25.3*	8.6	15.7	34.9	246.2
6	-14.4	*25.1*	9.7	17.0	36.9	254.5
8	-13.3	*24.8*	10.8	18.4	38.9	263.0
10	-12.2	*24.5*	12.0	19.7	41.0	271.8
12	-11.1	*24.2*	13.2	21.1	43.2	280.7
14	-10.0	*23.9*	14.4	22.6	45.4	289.9
16	-8.9	*23.5*	15.7	24.1	47.7	299.2
18	-7.8	*23.2*	17.1	25.7	50.0	*308.8*

SATURATION CHART

Italics Indicate inches Mercury   Pressure numbers indicate psig

TEMP.		REFRIGERANT NO. & CYL. COLOR CODE				
F°	C°	Silver 717	Orange 11	White 12	Green 22	Purple 113
-06	-21.1	11.6	*25.5*	6.3	19.3	*28.5*
-04	-20.0	12.9	*25.3*	7.2	20.8	*28.4*
-02	-18.9	14.3	*25.0*	8.2	22.4	*28.3*
0	-17.8	15.7	*24.7*	9.2	24.0	*28.2*
2	-16.7	17.2	*24.4*	10.2	25.6	*28.1*
4	-15.6	18.8	*24.1*	11.2	27.3	*28.0*
6	-14.4	20.4	*23.8*	12.3	29.1	*27.9*
8	-13.3	22.1	*23.4*	13.5	30.9	*27.7*
10	-12.2	23.8	*23.0*	14.6	32.8	*27.6*
12	-11.1	25.6	*22.7*	15.8	34.7	*27.5*
14	-10.0	27.5	*22.3*	17.1	36.7	*27.3*
16	-8.9	29.4	*21.9*	18.4	38.7	*27.1*
18	-7.8	31.4	*21.5*	19.7	40.9	*27.0*

F°	C°	R123	Blue R134a	Yellow 500	Orchid 502	Aqua 503
-06	-21.1	*26.5*	3.7	9.9	25.8	207.3
-04	-20.0	*26.3*	4.6	11.0	27.5	214.7
-02	-18.9	*26.1*	5.5	12.1	29.3	222.3
0	-17.8	*25.8*	6.5	13.3	31.1	230.0
2	-16.7	*25.6*	7.5	14.5	33.0	238.0
4	-15.6	*25.3*	8.6	15.7	34.9	246.2
6	-14.4	*25.1*	9.7	17.0	36.9	254.5
8	-13.3	*24.8*	10.8	18.4	38.9	263.0
10	-12.2	*24.5*	12.0	19.7	41.0	271.8
12	-11.1	*24.2*	13.2	21.1	43.2	280.7
14	-10.0	*23.9*	14.4	22.6	45.4	289.9
16	-8.9	*23.5*	15.7	24.1	47.7	299.2
18	-7.8	*23.2*	17.1	25.7	50.0	*308.8*

SATURATION CHART

Italics Indicate inches Mercury   Pressure numbers indicate psig

TEMP.		REFRIGERANT NO. & CYL. COLOR CODE				
F°	C°	Silver 717	Orange 11	White 12	Green 22	Purple 113
20	-6.7	**33.5**	*21.1*	**21.0**	**43.0**	*26.8*
22	-5.6	**35.7**	*20.6*	**22.4**	**45.3**	*26.6*
24	-4.4	**37.9**	*20.1*	**23.9**	**47.6**	*26.4*
26	-3.3	**40.2**	*19.7*	**25.4**	**50.0**	*26.2*
28	-2.2	**42.6**	*19.1*	**26.9**	**52.4**	*26.0*
30	-1.1	**45.0**	*18.6*	**28.5**	**54.9**	*25.8*
32	0.0	**47.6**	*18.1*	**30.1**	**57.5**	*25.6*
34	1.1	**50.2**	*17.5*	**31.7**	**60.1**	*25.3*
36	2.2	**52.9**	*16.9*	**33.4**	**62.9**	*25.1*
38	3.3	**55.7**	*16.3*	**35.2**	**65.6**	*24.8*
40	4.4	**58.6**	*15.6*	**37.0**	**68.5**	*24.5*
42	5.6	**61.6**	*15.0*	**38.8**	**71.5**	*24.2*
44	6.7	**64.7**	*14.1*	**40.7**	**74.5**	*23.9*

F°	C°	R123	Blue R134a	Yellow 500	Orchid 502	Aqua 503
20	-6.7	*22.8*	**18.4**	**27.3**	**52.5**	**318.5**
22	-5.6	*22.4*	**19.9**	**28.9**	**55.0**	**328.5**
24	-4.4	*22.0*	**21.4**	**30.6**	**57.5**	**338.7**
26	-3.3	*21.6*	**22.9**	**32.4**	**60.1**	**349.1**
28	-2.2	*21.2*	**24.5**	**34.2**	**62.8**	**359.7**
30	-1.1	*20.7*	**26.1**	**36.0**	**65.6**	**370.6**
32	0.0	*20.2*	**27.8**	**37.9**	**68.4**	**381.7**
34	1.1	*19.7*	**29.5**	**39.9**	**71.3**	**393.0**
36	2.2	*19.2*	**31.3**	**41.9**	**74.3**	**404.5**
38	3.3	*18.7*	**33.1**	**43.9**	**77.4**	**416.2**
40	4.4	*18.1*	**35.0**	**46.1**	**80.5**	**428.2**
42	5.6	*17.5*	**37.0**	**48.2**	**83.8**	**440.5**
44	6.7	*16.9*	**39.0**	**50.5**	**87.0**	**452.9**

SATURATION CHART

Italics Indicate inches Mercury   Pressure numbers indicate psig

TEMP.		REFRIGERANT NO. & CYL. COLOR CODE				
F°	C°	Silver 717	Orange 11	White 12	Green 22	Purple 113
46	7.8	67.9	_13.6_	42.7	77.6	_23.6_
48	8.9	71.1	_12.8_	44.7	80.8	_23.3_
50	10.0	74.5	_12.0_	46.7	84.0	_22.9_
55	12.8	83.4	_10.0_	52.0	92.5	_22.1_
60	15.6	92.9	_7.8_	57.7	101.6	_21.0_
65	18.3	103.1	_5.4_	63.8	111.2	_19.9_
70	21.1	114.1	_2.8_	70.2	121.4	_18.7_
75	23.9	125.8	0.0	77.0	132.2	_17.3_
80	26.7	138.3	1.5	84.2	143.6	_15.9_
85	29.4	151.7	3.2	91.8	155.6	_14.3_
90	32.2	165.9	4.9	99.8	168.4	_12.5_
95	35.0	181.1	6.8	108.3	181.8	_10.6_
100	37.8	197.2	8.8	117.2	195.9	_8.6_

F°	C°	R123	Blue R134a	Yellow 500	Orchid 502	Aqua 503
46	7.8	_16.3_	41.1	52.8	90.4	465.6
48	8.9	_15.6_	43.2	55.1	93.8	478.5
50	10.0	_15.0_	45.4	57.6	97.4	491.7
55	12.8	_13.1_	51.2	64.1	106.6	517.3
60	15.6	_11.2_	57.4	71.0	116.4	551.8
65	18.3	_9.0_	64.0	78.1	125.8	598.7
70	21.1	_6.6_	71.1	85.8	136.6	
75	23.9	_4.1_	78.6	93.9	147.9	
80	26.7	_1.3_	86.7	102.5	159.9	
85	29.4	0.9	95.2	111.5	172.5	
90	32.2	2.5	104.3	121.2	185.8	
95	35.0	4.2	113.9	131.3	199.7	
100	37.8	_6.1_	124.1	141.9	214.4	

SATURATION CHART

Italics Indicate inches Mercury   Pressure numbers indicate psig

TEMP.		REFRIGERANT NO. & CYL. COLOR CODE				
F°	C°	Silver 717	Orange 11	White 12	Green 22	Purple 113
105	30.6	214.2	11.1	126.6	210.7	*6.4*
110	43.3	232.3	13.4	136.4	226.3	*4.0*
115	46.1	251.5	15.9	146.8	242.7	*1.4*
120	48.8	271.7	18.5	157.7	259.6	0.7
125	51.7	293.1	21.3	169.1	277.9	2.2
130	54.4		24.3	181.0	296.8	3.7
135	57.2		27.4	193.5	316.5	5.4
140	60.0		30.8	206.6	337.2	7.2
145	62.8		34.4	220.3	358.8	9.2
150	65.6		38.2	234.6	381.5	11.2

F°	C°	R123	Blue R134a	Yellow 500	Orchid 502	Aqua 503
105	30.6	8.1	143.9	153.1	229.7	
110	43.3	10.2	146.3	164.9	245.8	
115	46.1	12.6	158.4	177.4	266.1	
120	48.8	15.0	171.1	190.3	280.3	
125	51.7	17.7	184.5	204.0	298.7	
130	54.4	20.5	198.7	218.2	318.0	
135	57.2	23.5	214.5	233.2	338.1	
140	60.0	26.7	229.2	248.8	359.2	
145	62.8	30.2	245.6	263.7	381.1	
150	65.6	33.8	262.8	280.7	404.0	

PSIG	R-401A (MP39) Liquid°F	Vapor°F	R-402A (HP80) Liquid°F	Vapor°F	R-404A (HP62) Liquid°F	Vapor°F
5" hg	-23		-59		-57	
4" hg	-22		-58		-56	
3" hg	-20		-56		-54	
2" hg	-19		-55		-53	
1" hg	-17		-54		-52	
0	-16		-53		-51	
1	-13		-50		-48	
2	-11		-48		-46	
3	-9		-45		-43	
4	-6		-43		-41	
5	-4		-41		-39	
6	-2		-39		37	
7	0		-37		-35	

PSIG	R-401A (MP39)		R-402A (HP80)		R-404A (HP62)	
	Liquid °F	Vapor °F	Liquid °F	Vapor °F	Liquid °F	Vapor °F
8	2		-36		-33	
9	4		-34		-32	
10	6		-32		-30	
11	8		-30		-28	
12	9		-29		-27	
13	11		-27		-25	
14	13		-26		-23	
15	14		-24		-22	
16	16		-23		-20	
17	17		-21		-19	
18	19		-20		-18	
19	20		-19		-16	
20	21		-17		15	

PSIG	R-401A (MP39)		R-402A (HP80)		R-404A (HP62)	
	Liquid °F	Vapor °F	Liquid °F	Vapor °F	Liquid °F	Vapor °F
21	23		-16		-14	
22	24		-15		-12	
23	25		-14		-11	
24	27		-12		-10	
25	28		-11		-9	
26	29		-10		-8	
27	30		-9		-6	
28	32		-8		-5	
29	33		-7		-4	
30	34		-6		-3	
31	35		-5		-2	
32	36		-4		-1	
33	37		-2		0	

PSIG	R-401A (MP39)		R-402A (HP80)		R-404A (HP62)	
	Liquid °F	Vapor °F	Liquid °F	Vapor °F	Liquid °F	Vapor °F
24	27		-12		-10	
25	28		-11		-9	
26	29		-10		-8	
27	30		-9		-6	
28	32	**BUBBLE POINT ⇒**	-8		-5	
29	33		-7		-4	
30	34		-6		-3	
31	35		-5		-2	
32	36		-4		-1	
33	37		-2		0	
34	38		-1		1	
35	39		0		2	
36	40	30	0		3	

PSIG	R-401A (MP39) Liquid °F	Vapor °F	R-402A (HP80) Liquid °F	Vapor °F	R-404A (HP62) Liquid °F	Vapor °F
37	42	31	1		4	
38	43	32	2		5	
39	44	33	3		6	
40	45	34	4		7	
42	46	36	6		8	
44	48	38	8		10	
46	50	40	10		12	
48	⇦ DEW POINT	42	11		14	
50		44	13		16	
52		45	14		17	
54		47	16		19	
56		49	18		20	
58		50	19		22	

PSIG	R-401A (MP39) Liquid°F	R-401A (MP39) Vapor°F	R-402A (HP80) Liquid°F	R-402A (HP80) Vapor°F	R-404A (HP62) Liquid°F	R-404A (HP62) Vapor°F
60		52	20		23	
62		53	22		25	
64		55	23		26	
66		56	25		27	
68		58	26		29	
70		59	27		30	29
72		61	29		32	31
74		62	30		33	32
76		64	31		34	33
78		65	32	30	35	34
80		66	34	31	37	36
85		69	37	34	40	39
90		73	40	37	42	42

BUBBLE POINT ⇨ (R-402A Vapor°F, beginning at 78 PSIG)

BUBBLE POINT ⇨ (R-404A Vapor°F, beginning at 70 PSIG)

PSIG	R-401A (MP39)		R-402A (HP80)		R-404A (HP62)	
	Liquid °F	Vapor °F	Liquid °F	Vapor °F	Liquid °F	Vapor °F
95		76	42	40	45	44
100		78	45	43	48	47
105		81	48	45		50
110		84	50	48		52
115		87		50		55
120		89		53		57
125		92		55		59
130		94		57		62
135		96		60		64
140		99		62		66
145		101		64		68
150		103		66		70
155		105		68		72

⇐ DEW POINT (R-402A)

⇐ DEW POINT (R-404A)

PSIG	R-401A (MP39)		R-402A (HP80)		R-404A (HP62)	
	Liquid °F	Vapor °F	Liquid °F	Vapor °F	Liquid °F	Vapor °F
160		108		70		74
165		110		72		76
170		112		74		78
175		114		75		80
180		116		77		82
185		117		79		83
190		119		81		85
195		121		82		87
200		123		84		88
205		125		86		90
210		127		87		92
220		130		91		95
230		133		94		98

14

PSIG	R-401A (MP39)		R-402A (HP80)		R-404A (HP62)	
	Liquid °F	Vapor °F	Liquid °F	Vapor °F	Liquid °F	Vapor °F
240		136		97		101
250		140		99		104
260		143		102		107
275		147		106		111
290		151		110		115
305		155		114		118
320		159		118		122
335		163		121		126
350		167		125		129
365		170		128		132

15

PSIG	R-407A		R-401 (MP66)		R-409A	
	Liquid °F	Vapor °F	Liquid °F	Vapor °F	Liquid °F	Vapor °F
5" hg	-45				-22	
4" hg	-43				-20	
3" hg	-42				-19	
2" hg	-41				-17	
1" hg	-39				-16	
0	-38		-30		-15	
1	-36				-12	
2	-33		-25		-9	
3	-31				-7	
4	-29		-20		-5	
5	-27				-2	
6	-25		-16		0	
7	-23				2	

PSIG	R-407A		R-401 (MP66)		R-409A	
	Liquid °F	Vapor °F	Liquid °F	Vapor °F	Liquid °F	Vapor °F
8	-21		-12	-2	4	
9	-20		-10	0	6	
10	-18		-8	2	8	
11	-16		-6	4	9	
12	-15		-4	6	11	
13	-13		-3	7	13	
14	-12		-1	9	14	
15	-10		0	10	16	
16	-9		2	12	17	
17	-8		3	13	19	
18	-6		5	15	20	
19	-5		6	16	22	
20	-4		8	18	23	

PSIG	R-407A		R-401 (MP66)		R-409A	
	Liquid °F	Vapor °F	Liquid °F	Vapor °F	Liquid °F	Vapor °F
21	-2		9	19	25	
22	-1		10	20	26	
23	0		11	21	27	
24	1		13	23	29	
25	2		14	24	30	
26	4		15	25	31	
27	5		16	26	32	
28	6		18	28	34	
29	7		19	29	35	
30	8		20	30	36	
31	9		21	31	37	
32	10		22	32	38	
33	11		23	33	39	

PSIG	R-407A		R-401 (MP66)		R-409A	
	Liquid °F	Vapor °F	Liquid °F	Vapor °F	Liquid °F	Vapor °F
34	12		25	34		BUBBLE POINT ⇒
35	13		26	36		
36	14		27	37		
37	15		28	38		
38	16		29	39	40	30
39	17		30	40	41	31
40	18		31	41	43	32
42	19		33	42	44	34
44	21		35	44	45	36
46	23		37	46	46	38
48	24		38	48	47	39
50	26		40	50	48	41
52	28		42	52	50 ⇐ DEW POINT	43

PSIG	R-407A		R-401 (MP66)		R-409A	
	Liquid °F	Vapor °F	Liquid °F	Vapor °F	Liquid °F	Vapor °F
54	29	BUBBLE POINT ⟹	44	54		45
56	31		45	55		46
58	32		46	56		48
60	33		48	58		50
62	35		49	59		51
64	36		51	61		53
66	38		52	62		54
68	39		53	63		56
70	40	30	55	65		57
72	41	31	56	66		58
74	43	32	58	68		60
76	44	34	59	69		61
78	45	35	61	71		63

PSIG	R-407A		R-401 (MP66)		R-409A	
	Liquid °F	Vapor °F	Liquid °F	Vapor °F	Liquid °F	Vapor °F
80	46	36	63	72		64
85	49	39	66	75		67
90		42	69	78		70
95		45	72	81		73
100		47	75	83		76
105		50	77	86		79
110		53	80	89		82
115		55	83	91		84
120		57	85	94		87
125		60	88	96		89
130		62	90	99		92
135		64	93	101		94
140		66	95	103		96

⇐ DEW POINT

PSIG	R-407A		R-401 (MP66)		R-409A	
	Liquid °F	Vapor °F	Liquid °F	Vapor °F	Liquid °F	Vapor °F
145		68	97	105		99
150		70	99	107		101
155		72	101	109		103
160		74	103	111		105
165		76	106	113		107
170		78	108	115		109
175		80	109	117		111
180		81	111	1119		113
185		83	113	121		115
190		85	115	123		117
195		87	117	125		119
200		88	119	126		121
205		90	121	128		123

PSIG	R-407A		R-401 (MP66)		R-409A	
	Liquid °F	Vapor °F	Liquid °F	Vapor °F	Liquid °F	Vapor °F
210		91	122	130		124
220		94	126	133		128
230		97	129	136		131
240		100	132	139		134
250		103	135	142		137
260		106	138	145		141
275		110	142	149		145
290		114	147	153		149
305		117	150	157		153
320		121	155	161		157
335		124	159	165		161
350		128	162	168		165
365		131	165	171		169

R-408A Saturated Vapor/Liquid
Temperature / Pressure Chart

Temp °F	PSIG	
	Vapor	Liquid
-40	2.7	2.3
-39	3.1	2.7
-38	3.6	3.2
-37	4.1	3.6
-36	4.5	4.1
-35	5.0	4.5
-34	5.5	5.1
-33	6.1	5.6
-32	6.6	6.1
-31	7.2	6.6
-30	7.7	7.2
-29	8.2	7.7
-28	8.8	8.2
-27	9.3	8.8
-26	9.9	9.3
-25	10.4	9.9
-24	11.1	10.4
-23	11.7	11.0
-22	12.3	11.5
-21	13.0	12.1

Temp °F	PSIG	
	Vapor	Liquid
-20	13.6	12.7
-19	14.3	13.3
-18	15.0	14.0
-17	15.6	14.6
-16	16.3	15.3
-15	17.0	15.9
-14	17.7	16.6
-13	18.5	17.3
-12	19.2	18.0
-11	20.0	18.7
-10	20.7	19.4
-9	21.5	20.2
-8	22.3	20.9
-7	23.1	21.7
-6	23.9	22.5
-5	24.7	23.2
-4	25.6	24.3
-3	26.5	25.3
-2	27.4	26.3
-1	28.3	27.3

R-408A Saturated Vapor/Liquid
Temperature / Pressure Chart Continued

Temp °F	PSIG	
	Vapor	Liquid
0	29.1	28.3
1	30.1	29.3
2	31.1	30.2
3	32.0	31.2
4	33.0	32.1
5	33.9	33.0
6	35.0	34.1
7	36.0	35.1
8	37.0	36.1
9	38.0	37.2
10	39.1	38.2
11	40.2	39.3
12	41.3	40.4
13	42.4	41.5
14	43.5	42.5
15	44.6	43.6
16	45.8	44.8
17	47.0	46.0
18	48.2	47.2
19	49.4	48.4

R-408A Saturated Vapor/Liquid
Temperature / Pressure Chart Continued

Temp °F	PSIG	
	Vapor	Liquid
20	50.6	49.6
21	51.9	50.8
22	53.1	52.1
23	54.4	53.3
24	57.0	55.8
25	58.4	57.2
26	59.8	58.6
27	61.1	60.0
28	62.5	61.4
29	63.9	62.7
30	63.9	62.7
31	65.3	64.2
32	66.8	65.6
33	68.3	67.1
34	69.8	68.6
35	71.2	70.0
36	72.8	71.6
37	74.4	73.2
38	76.0	74.7
39	77.6	76.3

R-408A Saturated Vapor/Liquid
Temperature / Pressure Chart Continued

Temp °F	PSIG	
	Vapor	Liquid
40	79.2	77.9
41	80.9	79.5
42	82.6	81.2
43	84.2	82.9
44	85.9	84.5
45	87.6	86.2
46	89.4	88.0
47	91.2	89.8
48	93.0	91.6
49	94.8	93.3
50	96.6	95.1
51	98.5	97.0
52	100.4	99.0
53	102.3	100.9
54	104.2	102.8
55	106.1	104.7
56	108.1	106.7
57	110.1	108.7
58	112.1	110.7
59	114.2	112.7

R-408A Saturated Vapor/Liquid
Temperature / Pressure Chart Continued

Temp °F	PSIG	
	Vapor	Liquid
60	116.2	114.7
61	118.4	116.9
62	120.5	119.0
63	122.7	121.2
64	124.9	123.4
65	127.1	125.5
66	129.3	127.8
67	131.6	130.0
68	133.9	132.3
69	136.1	134.5
70	138.4	136.7
71	140.8	139.1
72	143.3	141.6
73	145.7	144.0
74	148.1	146.4
75	150.6	148.8
76	153.1	151.4
77	155.6	153.9
78	158.2	156.5
79	160.7	159.1

R-408A Saturated Vapor/Liquid
Temperature / Pressure Chart Continued

Temp °F	PSIG	
	Vapor	Liquid
80	163.3	161.6
81	165.9	164.3
82	168.6	167.0
83	171.3	169.7
84	174.0	172.4
85	176.7	175.0
86	179.6	177.9
87	182.4	180.7
88	185.3	183.5
89	188.2	186.3
90	191.0	189.1
91	194.0	192.1
92	197.0	195.1
93	200.0	198.1
94	202.9	201.1
95	205.9	204.1
96	209.1	207.2
97	212.3	210.4
98	215.4	213.5
99	218.6	216.7

R-408A Saturated Vapor/Liquid
Temperature / Pressure Chart Continued

Temp °F	PSIG	
	Vapor	Liquid
100	221.8	219.8
101	225.1	223.1
102	228.4	226.4
103	231.7	229.7
104	235.0	233.0
105	238.3	236.3
106	241.8	239.8
107	245.2	243.2
108	248.7	246.7
109	252.2	250.1
110	255.7	253.6
111	259.3	257.3
112	262.9	260.9
113	26636	264.6
114	270.2	268.3
115	273.8	272.0
116	277.7	275.9
117	281.7	279.7
118	285.6	283.6
119	289.5	287.5

R-408A Saturated Vapor/Liquid
Temperature / Pressure Chart Continued

Temp °F	PSIG	
	Vapor	Liquid
120	293..4	219.4
121	297.5	295.5
122	301.6	299.5
123	305.6	303.6
124	309.7	307.7
125	313.8	311.7
126	317.8	315.8
127	321.9	319.9
128	326.0	32.9
129	330.0	328.0
130	334.1	332.1

HCFC 22 SATURATION PROPERTIES – TEMPERATURE TABLE

| TEMP. | PRESSURE | | ENTHALPY | | | ENTROPY | |
| °F | | | Btu/lb | | | Btu/(lb)(°R) | |
	PSIA	PSIG	LIQ- h_f	LA- h_{fg}	VAPOR h_g	LIQUID s_f	VAPOR s_g
10	47.464	32.768	13.104	92.338	105.442	0.02932	0.22592
11	48.423	33.727	13.376	92.162	105.538	0.02990	0.22570
12	49.396	34.700	13.648	91.986	105.633	0.03047	0.22548
13	50.384	35.688	13.920	91.808	105.728	0.03104	0.22527
14	51.387	36.691	14.193	91.630	105.823	0.03161	0.22505
15	52.405	37.709	14.466	91.451	105.917	0.03218	0.22484
16	53.438	38.742	14.739	91.272	106.011	0.03275	0.22463
17	54.487	39.791	15.013	91.091	106.105	0.03332	0.22442
18	55.551	40.855	15.288	90.910	106.198	0.03389	0.22421
19	56.631	41.935	15.562	90.728	106.290	0.03446	0.22400
20	57.727	43.031	15.837	90.545	106.383	0.03503	0.22379
21	58.839	44.143	16.113	90.362	106.475	0.03560	0.22358
22	59.967	45.271	16.389	90.178	106.566	0.03617	0.22338
23	61.111	46.415	16.665	89.993	106.657	0.03674	0.22318
24	62.272	47.576	16.942	89.807	106.748	0.03730	0.22297

HCFC 22 SATURATION PROPERTIES – TEMPERATURE TABLE

TEMP.	PRESSURE		ENTHALPY			ENTROPY	
°F			Btu/lb			Btu/(lb)(°R)	
	PSIA	PSIG	LIQUID	LATENT	VAPOR	LIQUID	VAPOR
			h_f	h_{fg}	h_g	s_f	s_g
25	63.450	48.754	17.219	89.620	106.839	0.03787	0.22277
26	64.644	49.948	17.496	89.433	106.928	0.03844	0.22257
27	65.855	51.159	17.774	89.244	107.018	0.03900	0.22237
28	67.083	52.387	18.052	89.055	107.107	0.03958	0.22217
29	68.328	53.632	18.330	88.865	107.196	0.04013	0.22198
30	69.591	54.895	18.609	88.674	107.284	0.04070	0.22178
31	70.871	56.175	18.889	88.483	107.372	0.04126	0.22158
32	72.169	57.473	19.169	88.290	107.459	0.04182	0.22139
33	73.485	58.789	19.449	88.097	107.546	0.04239	0.22119
34	74.818	60.122	19.729	87.903	107.632	0.04295	0.22100
35	76.170	61.474	20.010	87.708	107.719	0.04351	0.22081
36	77.540	62.844	20.292	87.512	107.804	0.04407	0.22062
37	78.929	64.233	20.574	87.316	107.889	0.04464	0.22043
38	80.336	65.640	20.856	87.118	107.974	0.04520	0.22024
39	81.761	67.065	21.138	86.920	108.058	0.04576	0.22005

HCFC 22 SATURATION PROPERTIES – TEMPERATURE TABLE

TEMP.	PRESSURE		ENTHALPY			ENTROPY	
°F			Btu/lb			Btu/(lb)(°R)	
	PSIA	PSIG	LIQUID	LATENT	VAPOR	LIQUID	VAPOR
			h_f	h_{fg}	h_g	s_f	s_g
40	83.206	68.510	21.422	86.720	108.142	0.04632	0.21986
41	84.670	69.974	21.705	86.520	108.225	0.04688	0.21968
42	86.153	71.457	21.989	86.319	108.308	0.04744	0.21949
43	87.655	72.959	22.273	86.117	108.390	0.04800	0.21931
44	89.177	74.481	22.558	85.914	108.472	0.04855	0.21912
45	90.719	76.023	22.843	85.710	108.553	0.04911	0.21894
46	92.280	77.584	23.129	85.506	108.634	0.04967	0.21876
47	93.861	79.165	23.415	85.300	108.715	0.05023	0.21858
48	95.463	80.767	23.701	85.094	108.795	0.05079	0.21839
49	97.085	82.389	23.988	84.886	108.874	0.05134	0.21821
50	98.727	84.031	24.275	84.678	108.953	0.05190	0.21803
51	100.39	85.69	24.563	84.468	109.031	0.05245	0.21785
52	102.07	87.38	24.851	84.258	109.109	0.05301	0.21768
53	103.78	89.08	25.139	84.047	109.186	0.05357	0.21750
54	105.50	90.81	25.429	83.834	109.263	0.05412	0.21732

35

HCFC 22 SATURATION PROPERTIES – TEMPERATURE TABLE

TEMP.	PRESSURE		ENTHALPY			ENTROPY	
°F			Btu/lb			Btu/(lb)(°R)	
	PSIA	PSIG	LIQUID	LATENT	VAPOR	LIQUID	VAPOR
			h_f	h_{fg}	h_g	s_f	s_g
55	107.25	92.56	25.718	83.621	109.339	0.05468	0.21714
56	109.02	94.32	26.008	83.407	109.415	0.05523	0.21697
57	110.81	96.11	26.298	83.191	109.490	0.05579	0.21679
58	112.62	97.93	26.589	82.975	109.564	0.05634	0.21662
59	114.46	99.76	26.880	82.758	109.638	0.05689	0.21644
60	116.31	101.62	27.172	82.540	109.712	0.05745	0.21627
61	118.19	103.49	27.464	82.320	109.785	0.05800	0.21610
62	120.09	105.39	27.757	82.100	109.857	0.05855	0.21592
63	122.01	107.32	28.050	81.878	109.929	0.05910	0.21575
64	123.96	109.26	28.344	81.656	110.000	0.05966	0.21558

R-12 PROPERTIES OF LIQUID AND SATURATED VAPOR

°F	PRESSURE		VOLUME VAPOR	DENSITY LIQUID	HEAT CONTENT BTU/LB.	
	psia	psig	cu . ft./lb.	cu . ft./lb.	liquid	vapor
-100	1.42	27.01*	22.16	100.15	-12.47	66.20
- 75	3.38	23.02*	9.92	97.93	- 7.31	69.00
- 50	7.11	15.43*	4.97	95.62	- 2.10	71.80
- 25	13.55	2.32*	2.73	93.20	3.17	74.56
- 15	17.14	2.45	2.19	92.20	5.30	75.65
- 10	19.18	4.49	1.97	91.70	6.37	76.20
- 5	21.42	6.73	1.78	91.18	7.44	76.73
0	23.85	9.15	1.61	90.66	8.52	77.27
5**	26.48	11.79	1.46	90.14	9.60	77.80

* Indicates inches of mercury (vacuum)

** Indicates "Standard Conditions." (5°F evaporating temperature and 86°F condensing temperature.)

R-12 PROPERTIES OF LIQUID AND SATURATED VAPOR

°F	PRESSURE		VOLUME VAPOR	DENSITY LIQUID	HEAT CONTENT BTU/LB.	
	psia	psig	cu . ft./lb.	lb./cu. ft.	liquid	vapor
10	29.34	14.64	1.32	89.61	10.68	78.34
25	39.31	24.61	1.00	87.98	13.96	79.90
50	61.39	46.70	.66	85.14	19.50	82.43
75	91.68	76.99	.44	82.09	25.20	84.82
86**	108.04	93.34	.38	80.67	27.77	85.82
100	131.86	117.16	.31	78.79	31.10	87.63
125	183.76	169.06	.22	75.15	37.28	88.97
150	249.31	234.61	.16	71.04	43.85	90.53
175	330.64	315.94	.11	66.20	51.03	91.48
200	430.09	415.39	.08	60.03	59.20	91.28

* Indicates inches of mercury (vacuum)

** Indicates "Standard Conditions." (5°F evaporating temperature and 86°F condensing temperature.)

R-22   PROPERTIES OF LIQUID AND SATURATED VAPOR

°F	PRESSURE		VOLUME VAPOR	DENSITY LIQUID	HEAT CONTENT BTU/LB.	
	psia	psig	cu . ft./lb.	lb./cu. ft.	liquid	vapor
-150	.27	29.37*	141.23	98.34	-25.97	87.52
-125	.89	28.12*	46.69	96.04	-20.33	90.43
-100	2.40	25.04*	18.43	93.77	-14.56	93.37
-75	5.61	18.50*	8.36	91.43	-8.64	96.29
-50	11.67	6.15*	4.22	89.00	-2.51	99.14
-25	22.09	7.39	2.33	86.78	3.83	101.88
-15	27.87	13.17	1.87	85.43	6.44	102.93
-10	31.16	16.47	1.68	84.90	7.75	103.46
5	34.75	20.06	1.52	84.37	9.08	103.97

*   Indicates inches of mercury (vacuum)

**   Indicates "Standard Conditions." (5°F evaporating temperature and 86°F condensing temperature.)

R-22 PROPERTIES OF LIQUID AND SATURATED VAPOR

°F	PRESSURE psia	VOLUME VAPOR psig	DENSITY LIQUID cu.ft./lb.	HEAT CONTENT BTU/LB. lb./cu. ft.	liquid	vapor
0	38.66	23.96	1.37	83.83	10.41	104.47
5**	42.89	28.19	1.24	83.28	11.75	104.96
10	47.46	32.77	1.13	82.72	13.10	105.44
25	63.45	48.75	.86	81.02	17.22	106.83
50	98.73	84.03	.56	78.03	24.28	108.95
75	146.91	132.22	.37	74.80	31.61	110.74
86**	172.87	158.17	.32	73.28	34.93	111.40
100	210.60	195.91	.26	71.24	39.67	112.11
125	292.60	277.92	.18	67.20	47.37	112.88
150	396.10	381.50	.12	62.40	56.14	112.73

*   Indicates inches of mercury (vacuum)

** Indicates "Standard Conditions." (5°F evaporating temperature and 86°F condensing temperature.)

R-134a PROPERTIES OF LIQUID AND SATURATED VAPOR

°F	PRESSURE		VOLUME VAPOR	DENSITY LIQUID	HEAT CONTENT BTU/LB.	
	psia	psig	cu . ft./lb.	lb./cu. ft.	liquid	vapor
- 50	5.6	15.43*	7.55	89.48	- 2.779	94.31
- 40	7.5	14.2*	5.72	88.49	00.00	95.82
- 30	9.9	9.56*	4.39	87.49	2.83	97.32
- 20	12.9	3.80*	3.41	86.47	5.71	98.81
- 15	14.7	0.00	3.02	85.96	7.17	99.55
- 10	16.7	1.95	2.69	85.44	8.64	100.28
- 5	18.8	4.18	2.39	84.91	10.13	101.01
0	21.2	6.55	2.14	84.38	11.63	101.74
5**	23.8	8.96	1.92	83.85	13.14	102.47

* Indicates inches of mercury (vacuum)

** Indicates "Standard Conditions." (5°F evaporating temperature and 86°F condensing temperature.)

R-134a PROPERTIES OF LIQUID AND SATURATED VAPOR

°F	PRESSURE		VOLUME VAPOR	DENSITY LIQUID	HEAT CONTENT BTU/LB.	
	psia	psig	cu . ft./lb.	lb./cu. ft.	liquid	vapor
10	26.7	12.00	1.72	83.31	14.66	103.19
15	29.8	15.13	1.55	82.76	16.19	103.90
20	33.1	18.74	1.40	82.21	17.74	104.61
25	36.8	21.82	1.26	81.65	19.30	105.31
50	60.1	45.62	0.78	78.75	27.27	108.73
75	93.3	78.42	0.50	75.63	35.57	111.95
86**	111.7	96.60	0.42	74.17	39.33	113.28
100	138.8	124.45	0.33	72.22	44.23	114.88
125	199.2	185.26	0.22	68.38	53.33	117.41
150	277.5	262.80	0.15	63.90	63.06	119.30

* Indicates inches of mercury (vacuum)

** Indicates "Standard Conditions." (5°F evaporating temperature and 86°F condensing temperature.)

R-502 PROPERTIES OF LIQUID AND SATURATED VAPOR

°F	PRESSURE		VOLUME VAPOR	DENSITY LIQUID	HEAT CONTENT BTU/LB.	
	psia	psig	cu . ft./lb.	lb./cu. ft.	liquid	vapor
-100	3.26	23.28*	10.46	97.86	-12.55	65.89
- 75	7.28	15.09*	4.96	95.24	- 7.59	68.92
- 50	14.60	.19*	2.59	92.51	- 2.25	71.93
- 25	26.82	12.13	1.47	89.68	3.50	74.87
- 15	33.49	18.80	1.19	88.50	5.91	76.02
- 10	37.26	22.56	1.07	87.90	7.13	76.58
- 5	41.35	26.66	.97	87.29	8.38	77.36
0	45.78	31.08	.88	86.68	9.63	77.69
5**	50.55	35.86	.80	86.06	10.91	78.24

*   Indicates inches of mercury (vacuum)

** Indicates "Standard Conditions." (5°F evaporating temperature and 86°F condensing temperature.)

43

R-502 PROPERTIES OF LIQUID AND SATURATED VAPOR

°F	PRESSURE		VOLUME VAPOR	DENSITY LIQUID	HEAT CONTENT BTU/LB.	
	psia	psig	cu . ft./lb.	lb./cu. ft.	liquid	vapor
10	55.70	41.00	.73	85.43	12.19	78.78
15	61.23	46.53	.67	84.80	13.49	79.31
25	73.50	58.81	.67	83.50	16.14	80.35
50	112.12	97.42	.56	80.06	22.98	82.80
75	163.82	149.13	.37	76.22	30.12	84.96
86**	191.28	176.59	.22	74.45	33.36	85.79
100	230.89	216.19	.17	71.97	37.56	86.71
125	316.60	301.36	.12	66.84	45.36	87.84
150	423.06	408.35	.08	60.09	53.85	87.76
175	559.41	544.72	.05	47.55	65.69	83.37

* Indicates inches of mercury (vacuum)

** Indicates "Standard Conditions." (5°F evaporating temperature and 86°F condensing temperature.)

THERMODYNAMIC PROPERTIES

	R-11	R-12	R-13	R-22	R-113	R-114	R-500	R-502	R-503
Properties at 1 Atmosphere:									
Freezing point, °F	-168	-252	-294	-256	-31	-137	-254		
Boiling point, °F	74.8	-21.6	-114.6	-41.4	117.6	38.4	28.3	-50.1	127.6
Condensation at 86°F:									
Specific Heat of Liquid, Btu/lb/cu ft	.209	.235	.247	.335	.218	.246	.290	.305	.290
Compressor Discharge Temp., °F	111	101	-1	128	86	86	105	99	14
Compressor Suction Temp., °F	5	5	-100	5	10	20	5	5	-100
Compressor Ratio	6.24	4.08	4.74	4.06	8.02	5.42	4.12	3.75	4.58

THERMODYNAMIC PROPERTIES

	R-11	R-12	R-13	R-22	R-113	R-114	R-500	R-502	R-503
Condensation at 86°F:									
Refrigerant Circulated per ton (lb/min)	2.96	4.00	4.30	2.89	3.73	4.64	3.30	4.38	3.72
Horsepower per ton	0.935	1.002	1.12	1.011	0.973	1.045	1.01	1.079	1.15
Coefficient of Performance	5.04	4.70	4.20	4.66	4.84	4.64	4.65	4.37	4.23
Evaporation at 5°F:									
Specific Volume cu ft/lb (suction gas)	12.27	1.46	1.55	1.25	27.38	4.34	1.50	0.82	1.32
Net Refrig. effect Btu/lb	66.8	50.0	46.3	70.0	53.7	44.7	60.6	44.9	55.4
Latent Heat of Vaporization. Btu/lb	83.5	68.2	52.1	93.2	70.6	61.1	82.5	68.9	72.1

46

STD. REFRG. DESIG.	CHEMICAL NAME	BOILING POINT °F	CHEMICAL FORMULA	MOLECULAR WEIGHT
11	Trichlorofluoromethane	74.8	CCl_3F	137.4
12	Dichlorodifluoromethane	-21.6	CCl_2F_2	120.9
13	Chlcrotrifluoromethane	-114.6	$CClF_3$	105.4
22	Chlorodifluoromethane	-41.4	$CHClF_2$	86.5
30	Methylene Chloride	105.2	CH_2Cl_2	84.9
40	Methyl Chloride	-10.8	CH_3Cl	50.5
50	Metnane	-259	CH_4	16.0
113	Trichlorotrifluoroethane	117.6	CCl_2FCClF_2	187.4
114	Dichlorotetrafluoroethane	38.4	$CClF_2CClF_2$	170.9

STD. REFRG. DESIG.	CHEMICAL NAME	BOILING POINT °F	CHEMICAL FORMULA	MOLECULAR WEIGHT
123	Dichlorotrifluoroethane	82.17	CCl_2HCF_3	152.93
134a	Tetrafluoroethane	-15.08	CF_3CFH_2	102.03
170	Ethane	-127.5	CH_3CH_3	30.0
290	Propane	-44.0	$CH_3CH_2CH_3$	44.0
500*	73.8% R-12 & 26.2% R-152a	-28.0	CCl_2F_2/CH_3CHF_2	99.3
502*	48.8% R-22 & 51.2% R-115	-50.1	$CHClF_2/CClF_2CF_3$	112.0
601	Isobutane	14.0	$CH(CH_3)_3$	58.1
717	Ammonia	-28.0	NH_3	17.0
718	Water	212.0	H_2O	18.0

* Denotes Azeotropic Mixture

LATENT HEAT		
SUBSTANCE	FREEZING OR MELTING BTU/LB.	LATENT HEAT OF VAPORIZATION OR CONDENSATION BTU/LB.
WATER	144	970.4
R-12		68.2
R-22		93.2
R-502		68.96
R-717 (Ammonia)		565.0

Refrigerant	EVAPORATING TEMPERATURE Degrees Fahrenheit				
	40	20	0	-20	-40
	Pressure Drop (PSIG)				
R-12 & R-500	2	1.5	1	0.75	0.5
R-22	3	2	1.5	1.0	0.75
R-502	3	2.5	1.75	1.25	1.0

REFRIGERANT APPLICATIONS

NUMBER & NAMES	CYLINDER COLOR CODE	CHEMICAL FORMULA	APPLICATION
R-11 (CFC)	ORANGE	CCl_3F (Trichloromonofluoromethane)	Large-scale air conditioning and refrigeration systems employing single or multi-stage centrifugal compressors. Also used as solvent and blowing agent for foams.
R-12 (CFC)	WHITE	CCl_2F_2 (Dichlorodifluoromethane)	Most popular of all refrigerants, widely used for air conditioning and refrigeration equipment. Also used as blowing agent for foams.
R-13 (CFC)	GRAY	$CClF_3$ (Chlorotrifluoromethane)	Used in the low stage of cascade systems to provide especially low evaporator temperatures.
R-22 (HCFC)	GREEN	$CHClF_2$ (Monochlorodifluoromethane)	Used for low temperature applications such as room air conditioners and home freezers, and for industrial equipment requiring very low temperatures.
R-113 (CFC)	PURPLE	$C_2Cl_3F_3$ (Trichlorotrifluoroethane)	Low pressure, may be used for centrifugal compressor systems.

REFRIGERANT APPLICATIONS

NUMBER & NAMES	CYLINDER COLOR CODE	CHEMICAL FORMULA	APPLICATION
R-114 (CFC)	DARK BLUE	$C_2Cl_2F_4$ (Dichlorotetrafluoroethane)	Used with centrifugal compressors when higher capacity or lower evaporator temperatures are needed. Also useful for rotary compressors in small appliances. Volume of vapor circulated per ton is about three times that for R-12.
R-123 (HCFC)	DARK GRAY	$CHCl_2CF_3$ (Dichlorotrifluoroethane)	As leading candidate in the next generation of blowing agents, HCFC-123 may offer effective solutions in such diverse applications as rigid board and foam systems insulation. Also may be used in centrifugal refrigeration equipment and in specialized solvent applications.
R-134a (HFC)	LIGHT AQUA	CF_3CH_2F (Tetrafluoroethane)	A hydrofluorocarbon with an ozone depletion potential of zero, HFC-134a holds great promise as a CFC substitute for a wide range of air conditioning and refrigeration systems in residential, commercial, and industrial applications.

REFRIGERANT APPLICATIONS

NUMBER & NAMES	CYLINDER COLOR CODE	CHEMICAL FORMULA	APPLICATION
R-500 (CFC)	YELLOW	CCl_2F_2/CH_3CHF_2 Azeotropic Mixture (Dichlorodifluoromethane/ Difluoroethane)	Similar to R-12. R-500 generally has higher discharge temperature range, increased re- frigerating capacity.
R-502 (CFC)	ORCHID	$CHClF_2/CClF_2CF_3$ Azeotropic Mixture (Monochlorodifluoromethane/ Monchloropentafluoroethane)	Low temperature commercial and industrial equipment. Used only with reciprocating compressors.

OZONE DEPLETION POTENTIAL	
REFRIGERANT COMPOUND	DEPLETION POTENTIAL
CFC-11	1.0
CFC-12	1.0
CFC-113	.8
CFC-602	.307
HCFC-22	.05
HCFC-123	.02
HCFC-124	.02
HCFC-141b	.10
HCFC-142b	.06
HFC-125	.00
HFC-134a	.00
HFC-152a	.00

REFRIGERATION SYSTEMS DATA

REFRIGERANT	APPROPRIATE LUBRICANT			
R-11			MO	
R-12	POE	AB	MO	
R-13	POE	AB	MO	
R-22	POE	AB	MO	
R-23	POE			
R-123	POE	AB	MO	
R-124	POE	AB		
R-125	POE			
R-134a	POE			PAG*
R-176			MO	
R-401A	POE	AB		
R-401B	POE	AB		
R-401C	POE	AB		
R-402A	POE	AB		
R-402B	POE	AB		
R-403B	POE	AB	MO	
R-404A	POE			
R-407A	POE			
R-407B	POE			
R-407C	POE			
R-410A	POE			
R-500	POE	AB	MO	
R-502	POE	AB	MO	
R-503	POE	AB	MO	
R-507	POE			
R-717			MO	

*PAG = polyalkylene glycol - R-134a automotive applications.
POE = polyol ester   AB = alkylbenzene   MO = mineral oil

AVERAGE COMPRESSOR CAPACITIES BTU/h

EVAPORATING TEMPERATURES °F

	-30°		-15°		+20°		+40°	
HP	110°	120°	110°	120°	110°	120°	110°	120°
2	5,200	4,500	9,100	8,200	18,000	16,800	22,800	21,200
3	9,000	8,300	14,300	13,200	22,100	20,800	36,300	34,200
5	14,200	12,500	24,800	22,400	41,700	39,300	62,400	58,500
7 ½	25,000	20,000	31,000	28,000	53,000	48,000	87,000	81,700
10	31,000	26,000	43,600	44,800	81,000	75,000	120,000	112,000
15	42,600	37,500	74,400	67,200	111,000	102,000	171,600	160,000
20	56,000	44,700	82,000	71,000	154,000	142,000	235,000	218,000
25	70,000	56,000	96,000	85,000	188,000	174,000	283,000	263,000
30	80,000	67,000	116,500	102,500	225,000	210,000	349,000	324,000
40	94,000	75,000	155,000	135,000	325,000	306,000	439,000	406,000
50	122,000	100,000	188,500	159,500	375,000	350,000	585,000	550,000
60	168,000	134,000	240,000	220,000	450,000	420,000	710,000	670,000
70	196,000	156,000	272,000	239,000	571,000	534,000	800,000	742,000

CONDENSING TEMPERATURE °F

R-12 RECOMMENDED REFRIGERANT LINE SIZES - SUCTION LINE

COMPRESSOR CAPACITY Btu/h	LENGTH OF RUN					
	15 Ft.	25 Ft.	35 Ft.	50 Ft.*	100 Ft*
18,500 - 20,000	5/8	5/8	5/8		
20,000 - 22,000	5/8	5/8	5/8		
22,000 - 24,000	5/8	5/8	5/8	3/4	3/4
24,000 - 34,000	5/8	5/8	3/4	3/4	3/4
38,000 - 40,000	3/4	3/4	3/4	7/8	7/8
40,000 - 44,000	3/4	7/8	7/8	7/8	7/8
44,000 - 51,000	7/8	7/8	7/8	7/8	7/8
53,000 - 66,000	7/8	7/8	7/8	1 1/8	1 1/8

Line sizes listed are outside tube dimensions, standard refrigeration tubing with .028 or .032 wall thickness.

Suggested sizes do not include consideration for additional pressure drop due to elbows, valves, etc.

Add 3 fluid ounces for each 10 ft of pipe over 35 ft

R-12 RECOMMENDED REFRIGERANT LINE SIZES - SUCTION LINE

COMPRESSOR CAPACITY Btu/h	LENGTH OF RUN					
	15 Ft.	25 Ft.	35 Ft.	50 Ft.*	100 Ft*	
18,5000 - 20,000	5/16	5/16	5/16			
20,000 - 22,000	5/16	5/16	5/16			
22,000 - 24,000	5/16	3/8	3/8	3/8	3/8	
24,000 - 34,000	5/16	3/8	3/8	3/8	3/8	
38,000 - 40,000	5/16	3/8	3/8	3/8	3/8	
40,000 - 44,000	3/8	3/8	3/8	3/8	3/8	
44,000 - 51,000	3/8	3/8	3/8	3/8	3/8	
53,000 - 66,000	1/2	1/2	1/2	1/2	1/2	

Line sizes listed are outside tube dimensions, standard refrigeration tubing with .028 or .032 wall thickness.

Suggested sizes do not include consideration for additional pressure drop due to elbows, valves, etc.

- Add 3 fluid ounces for each 10 ft of pipe over 35 ft

COMPRESSOR CAPACITY Btu/h	LENGTH OF RUN				
	15 Ft.	25 Ft.	35 Ft.	50 Ft.*	100 Ft.*
18,5000 - 20,000	5/16	3/8	3/8		
20,000 - 22,000	3/8	3/8	3/8		
22,000 - 24,000	3/8	3/8	3/8	1/2	1/2
24,000 - 34,000	3/8	3/8	1/2	1/2	1/2
38,000 - 40,000	3/8	1/2	1/2	1/2	1/2
40,000 - 44,000	3/8	1/2	1/2	1/2	1/2
44,000 - 51,000	3/8	1/2	1/2	1/2	5/8
53,000 - 66,000	1/2	1/2	5/8	5/8	3/4

Line sizes listed are outside tube dimensions, standard refrigeration tubing with .028 or .032 wall thickness.

Suggested sizes do not include consideration for additional pressure drop due to elbows, valves, etc.

- Add 3 fluid ounces for each 10 ft of pipe over 35 ft

RECOMMENDED CAPILLARY TUBE LENGTH AND DIAMETER

Compressor Capacity Btu/h	Condenser Type	Normal Evaporating Temperature			
		-10 to +5	+5 to +20	+20 to +35	+35 to
		R-12 LOW TEMPERATURE			
200 - 300	Static/fan	16' - .026"	10' - .026"		
300 - 400	Static/fan	12' - .026"	12' - .031"		
400 - 700	Static	12' - .031"	12' - .036"		
	Fan	10' - .031"	10' - .036"		
700 - 1100	Static	12' - .036"			
	Fan	10' - .036"			
1100 - 1300	Static	10' - .036"			
	Fan	8' - .036"			
1300 - 1700	Static	12' - .042"			
	Fan	10' - .042"			
1700 - 2000	Static	12' - .049"			
	Fan	10' - .042"			
2000 - 3000	Fan	10' - .054"	15' - .059"		

RECOMMENDED CAPILLARY TUBE LENGTH AND DIAMETER

Compressor Capacity Btu/h	Condenser Type	Normal Evaporating Temperature			
		−10 to +5	+5 to +20	+20 to +35	+35 to +50
R-12 LOW TEMPERATURE					
3000 − 4000	Fan	10' − .059"	12' − .064"		
4000 − 4500	Fan	12' − .064"	12' − .070"		
4500 − 5000	Fan	10' − .070"	12' − .080"		
5000 − 7000	Fan	10' − .059" (2 pcs.)	12' − .064" (2 pcs.)		
7000 − 9000	Fan	10' − .064" (2 pcs.)	10' − .070" (2 pcs.)		
9000 − 12,000	Fan	10' − .070" (2 pcs.)	12' − .080" (2 pcs.)		
12,000 − 15,000	Fan	10' − .070" (2 pcs.)	12' − .080" (2 pcs.)		

RECOMMENDED CAPILLARY TUBE LENGTH AND DIAMETER

R-12 LOW TEMPERATURE

Compressor Capacity Btu/h	Condenser Type	Normal Evaporating Temperature			
		-10 to +5	+5 to +20	+20 to +35	+35 to +50
1000 - 2000	Fan	10' - .036"	12' - .042"		
2000 - 3000	Fan	12' - .042"	15' - .049"		
3000 - 4000	Fan	10' - .054"	15' - .059"		
4000 - 5000	Fan	10' - .064"	15' - .070"		

RECOMMENDED CAPILLARY TUBE LENGTH AND DIAMETER

Compressor Capacity Btu/h	Condenser Type	Normal Evaporating Temperature			
		-10 to +5	+5 to +20	+20 to +35	+35 to +50
R-12 LOW TEMPERATURE					
1400 - 1600	Fan		12' - .036"	8' - .036"	8' - .042"
1600 - 1800	Fan		10' - .036"	12' - .042"	
1800 - 2500	Fan		12' - .042"	12' - .049"	8' - .049"
2500 - 3500	Fan		10' - .042"	10' - .049"	
3500 - 4000	Fan		12' - .049"	10' - .054"	
4000 - 5000	Fan		10' - .054"	10' - .059"	
5000 - 6000	Fan		12' - .059	12' - .064"	
6000 - 7000	Fan		10' - .059"	10' - .064"	
7000 - 10.000	Fan		12' - .070" 12' - .054" (2 pcs.)	12' - .080" 10' - .059" (2 pcs.)	
10,000 - 13,000	Fan		12' - .059" (2 pcs.)	10' - .064 (2 pcs.)	

| RECTANGULAR DUCTS WIDE SIDE | COMMERCIAL | | RESIDENTIAL |
	SHEET STEEL GALVANIZED (GAUGE)	ALUMINUM	SHEET STEEL GALVANIZED (GAUGE)
up to 12"	26	.020"	28
13" - 23"	24	.025"	26
24" - 30"	24	.025"	24
31" - 42"	22	.032"	
55" - 60"	20	.040"	
61" - 84"	20	.040"	
85" - 96"	18	.050"	
over 96"	18	.050"	

LOW SIDE PRESSURE MOTOR CONTROL SETTINGS	R-12		R-22		R-502	
* = inches vacuum	OUT	IN	OUT	IN	OUT	IN
Freezer - Open Type	7*	5	4	17	9	23
Freezer - Closed Type	1	8	11	22	17	29
Ice Cube Maker	4	17	16	37	22	47
Sweet Water Bath - Soda Fountain	21	29	43	56	52	67
Showcase - Frost Cycle	10	25	25	50	32	60
Showcase - Defrost Cycle	18	34	39	64	49	76
Beer, Water, Milk Cooler	19	29	40	56	47	67
Walk-in Cooler - Defrost Cycle	12	35	27	66	35	77
Vegetable Display - Defrost Cycle	11	35	27	66	35	77
Eutectic Brine Tank - Ice Cream Truck	1	4	11	16	17	22
Reach-in Cooler - Defrost Cycle	18	36	39	68	47	79
Beer Coolers - Blower Dry Type	15	34	33	64	42	76
Beer Coolers - Bare Pipe Dry Frost	12	27	29	53	37	64
Instantaneous Beer Coolers	12	29	29	56	37	67
Retail Florist Box - Blower Coil	26	42	51	77	61	88

HEAD PRESSURE FOR WATER COOLED CONDENSERS

	INLET WATER TEMPERATURE °F										
	50	55	60	65	70	75	80	85	90	95	100
REFRIGERANT	HEAD PRESSURE (PSIG)										
R-12	56	62	68	74	80	87	93	101	108	117	125
R-134a	55	61	68	76	84	92	101	110	120	131	142
R-22	95	104	113	123	133	144	155	158	180	194	208
R-717	98	108	119	130	141	152	164	177	191	205	220
R-500	71	78	86	94	103	112	121	131	142	153	165
R-502	116	126	137	148	160	173	186	200	214	230	246
R-404a	120	131	143	155	168	182	196	212	228	245	263

SUMMER DESIGN TEMPERATURE

STATE	DESIGN DRY BULB °F	°C	STATE	DESIGN DRY BULB °F	°C
ALABAMA	95	29	MASSACHUSETTS	90	32
ALASKA	74	23	MICHIGAN	88	31
ARIZONA	105	41	MINNESOTA	90	32
ARKANSAS	98	37	MISSISSIPPI	97	32
CALIFORNIA			MISSOURI	98	37
LOWER	86	30	MONTANA	88	31
MIDDLE	94	34	NEBRASKA	97	36
UPPER	83	28	NEVADA	95	35
COLORADO	92	33	NEW HAMPSHIRE	90	32
CONNECTICUT	88	31	NEW JERSEY	92	33
DELAWARE	93	34	NEW MEXICO	95	35
D.C.	94	34	NEW YORK	90	32

State			State		
FLORIDA			NORTH CAROLINA	95	35
UPPER	96	36	NORTH DAKOTA	93	34
LOWER	93	34	OHIO	90	32
GEORGIA	95	35	OKLAHOMA	102	39
HAWAII	87	31	OREGON	90	32
IDAHO	94	34	PENNSYLVANIA	92	33
ILLINOIS			RHODE ISLAND	87	31
UPPER	96	36	SOUTH CAROLINA	95	35
LOWER	93	34	SOUTH DAKOTA	95	35
INDIANA	95	35	TENNESSEE	96	36
IOWA	95	35	TEXAS	101	38
KANSAS			UTAH	95	35
UPPER	97	36	VERMONT	87	31
LOWER	100	38	VIRGINIA	95	35
KENTUCKY	95	35	WASHINGTON	90	32
LOUISIANA	98	37	WEST VIRGINIA	94	34
MAINE	88	31	WISCONSIN	90	32
MARYLAND	94	34	WYOMING	90	32

A/C & R
TROUBLESHOOTING

LOW HEAD PRESSURE

Possible Cause	Remedy
Low charge, cannot flood condenser.	Add refrigerant.
Low pressure setting on valve.	Increase setting.
Valve fails to close due to foreign material stuck in valve.	Open valve wide open to pass material. If unsuccessful, replace valve.
Valve will not adjust.	Replace valve.
Valve fails to close due to loss of element charge.	Replace valve.
Valve fails to open.	Replace valve.

HIGH HEAD PRESSURE

Possible Cause	Remedy
Dirty condenser.	Clean condenser.
System overcharged.	Purge until proper head pressure is obtained.
Undersized receiver.	Check receiver capacity.
Air in system.	Purge air from system.
Valve fails to adjust or open.	Replace valve.
Bypassing hot gas when not required.	Replace valve.

HERMETIC SYSTEM TROUBLESHOOTING

COMPLAINT	POSSIBLE CAUSE	REPAIR
1. Compressor will not start - no hum	A. Line disconnect switch open B. Fuse blown C. Overload protector tripped D. Control stuck in open position E. Control open because of cold location F. Incorrect wiring or loose connection	A. Close disconnect switch B. Replace fuse C. Refer to electrical diagram D. Repair or replace control E. Relocate control F. Refer to electrical diagram

HERMETIC SYSTEM TROUBLESHOOTING

COMPLAINT	POSSIBLE CAUSE	REPAIR
2. Compressor will not start - hums	A. Improper wiring B. Low voltage to unit C. Start capacitor defective D. Relay failing to close E. Open or shorted motor winding F. Mechanical failure in compressor	A. Refer to wiring diagram B. Correct voltage supply C. Replace start capacitor D. Check relay connections; if ok, replace relay E. Repair winding F. Repair compressor

HERMETIC SYSTEM TROUBLESHOOTING

COMPLAINT	POSSIBLE CAUSE	REPAIR
3. Compressor will not start - hums, then trips on overload protector	A. Incorrect wiring B. Low voltage to unit C. Relay failing to open D. Run capacitor defective E. Excessively high discharge pressure F. Open or shorted compressor winding G. Mechanical failure in compressor	A. Refer to wiring diagram B. Correct voltage supply C. Check wiring to relay; if ok, replace relay D. Replace capacitor E. Check discharge service valve; overcharged system F. Repair winding G. Repair compressor

HERMETIC SYSTEM TROUBLESHOOTING

COMPLAINT	POSSIBLE CAUSE	REPAIR
4. Compressor starts and runs, but short cycles on overload protector	A. Additional current through overload B. Low voltage to unit C. Defective overload protector D. Run capacitor defective E. Excessive discharge pressure	A. Refer to wiring diagram, check for added fans, pumps, motors, etc., Connected to wrong side of overload. B. Correct voltage supply C. Check current, replace overload D. Replace run capacitor E. Check condenser air flow, fan, restrictions

HERMETIC SYSTEM TROUBLESHOOTING

COMPLAINT	POSSIBLE CAUSE	REPAIR
5. Unit operates long or continuously	A. Undercharged system B. Control contacts stuck closed C. Refrigerated space has excessive load or poor insulation D. Undersized system E. Evaporator coil iced	A. Repair leak, adjust charge B. Clean contacts or replace control C. Determine fault and correct D. Replace with larger system E. Defrost

HERMETIC SYSTEM TROUBLESHOOTING

COMPLAINT	POSSIBLE CAUSE	REPAIR

6. Start capacitor open or shorted	A. Relay contacts not operating properly
	B. Long start cycle due to:
	  1. Low voltage to unit
	  2. Improper relay
	  3. Start load too high
	C. Excessive short cycling
		A. Clean contacts or replace relay
		B. 1. Correct voltage supply
		  2. Replace relay
		  3. Correct by installing pump down system
		C. Troubleshoot short cycling

HERMETIC SYSTEM TROUBLESHOOTING

COMPLAINT	POSSIBLE CAUSE	REPAIR
7. Run capacitor open or shorted	A. Improper capacitor B. Excessively high line voltage (over 110% of rated max)	A. Replace with correct size capacitor B. Correct voltage supply

HERMETIC SYSTEM TROUBLESHOOTING

COMPLAINT	POSSIBLE CAUSE	REPAIR
8. Relay defective or burned out	A. Incorrect relay B. Incorrect mounting angle C. Line voltage too high or too low D. Excessive short cycling E. Relay mounting loose or vibrating F. Incorrect run capacitor	A. Replace with correct relay B. Mount relay in correct position C. Correct voltage supply D. Troubleshoot short cycling E. Mount securely F. Replace with correct capacitor

HERMETIC SYSTEM TROUBLESHOOTING

COMPLAINT	POSSIBLE CAUSE	REPAIR
9. Space temp too high	A. Control setting too high B. Poor air circulation	A. Reset control B. Improve air movement

HERMETIC SYSTEM TROUBLESHOOTING

COMPLAINT	POSSIBLE CAUSE	REPAIR
10. Suction line frosted or sweating	A. Evaporator fan not running B. Refrigerant overcharge	A. Repair fan B. Adjust charge

HERMETIC SYSTEM TROUBLESHOOTING

COMPLAINT	POSSIBLE CAUSE	REPAIR
11. Liquid line frosted or sweating	A. Restriction in filter / drier	A. Replace drier

HERMETIC SYSTEM TROUBLESHOOTING

COMPLAINT	POSSIBLE CAUSE	REPAIR
12. Unit runs but short cycles	A. Defective overload protector	A. Check current, replace overload
	B. Thermostat	B. Differential set too close
	C. High pressure cut out due to:	C. 1. Correct airflow to condenser
	1. Insufficient airflow	2. Repair leak, adjust charge
	2. Overcharge	3. Purge system
	3. Air in system	
	D. Low pressure cut out due to:	D. 1. Leak test, adjust system
	1. Undercharge	2. Replace refrigerant control
	2. Restriction in refrigerant control	

ELECTRICAL

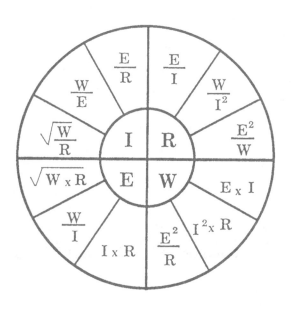

I = Intensity (Amperes)
R = Resistance (Ohms)
E = Electromotive Force (Voltage)
W = Watts

Electrical Symbols Used On Schematics

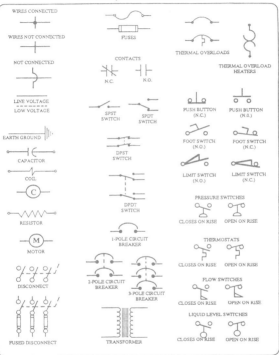

WIRES CONNECTED

WIRES NOT CONNECTED

NOT CONNECTED

LINE VOLTAGE
LOW VOLTAGE

EARTH GROUND

CAPACITOR

COIL

C

RESISTOR

MOTOR

DISCONNECT

FUSED DISCONNECT

FUSES

CONTACTS

N.C.   N.O.

SPST
SWITCH   SPDT
SWITCH

DPST
SWITCH

DPDT
SWITCH

1-POLE CIRCUIT
BREAKER

2-POLE CIRCUIT
BREAKER

3-POLE CIRCUIT
BREAKER

TRANSFORMER

THERMAL OVERLOADS

THERMAL OVERLOAD
HEATERS

PUSH BUTTON
(N.C.)   PUSH BUTTON
(N.O.)

FOOT SWITCH
(N.O.)   FOOT SWITCH
(N.C.)

LIMIT SWITCH
(N.O.)   LIMIT SWITCH
(N.C.)

PRESSURE SWITCHES

CLOSES ON RISE   OPEN ON RISE

THERMOSTATS

CLOSES ON RISE   OPEN ON RISE

FLOW SWITCHES

CLOSES ON RISE   OPEN ON RISE

LIQUID LEVEL SWITCHES

CLOSES ON RISE   OPEN ON RISE

| Electrical Engineering Units and Constants |||||
| Symbols and Units |||||

Quantity	Symbol	Unit	Unit Abbreviation
charge	Q	coulomb	C
current	I	ampere	A
voltage, potential difference	V	volt	V
electromotive force	ε	volt	V
resistance	Ω	ohm	R
conductance	G	mho	A/V or mho (S)
reactance	X	ohm	ohm
susceptance	B	mho	A/V or mho
impedance	Z	ohm	ohm
admittance	Y	mho	A/V or mho
capacitance	C	farad	F
inductance	L	henry	H
energy, work	W	joule	J
power	P	watt	W
resistivity	ρ	ohm-meter	Ωm
conductivity		mho per meter	mho / m
electric displacement	D	coulomb per sq. meter	C / m^2
electric field strength	E	volt per meter	V / m
magnetic flux	φ	weber	Wb
magnetic flux density	B	tesla	T
reluctance	ℜ	ampere per weber	A / Wb

COLOR CODE FOR RESISTORS			
COLOR	BAND A & B	BAND C	BAND D
BLACK	0	1	
BROWN	1	10	
RED	2	100	
ORANGE	3	1,000	
YELLOW	4	10,000	
GREEN	5	100,000	
BLUE	6	1,000,000	
VIOLET	7		
GRAY	8		
WHITE	9		
SILVER		0.01	± 10 %
GOLD		0.1	± 5 %
NO COLOR			± 20 %

CONDUCTANCE AND RESISTANCE OF VARIOUS METALS

METAL	CONDUCTANCE	RESISTANCE
Silver	2207	1.0
Copper	2030	1.09
Gold	1471	1.50
Aluminum	1277	1.73
Tungsten	634	3.48
Brass	461	4.79
Iron	304	7.26
Nickel	263	8.39
Steel	235	9.39
Nichrome	34	64.91

INSULATED CONDUCTOR APPLICATIONS

TYPE OF INSULATION	LETTER CODE	MAX. TEMP.	APPLICATION
Asbestos	A	392F	Dry locations only
Asbestos and Varnished Cambric	AVA	230F	Dry locations only
Heat Resistant Rubber	RH	167F	Dry and damp locations
Thermoplastic	T	140F	Dry locations
Moisture Resistant Thermoplastic	TW	140F	Dry and wet locations
Heat Resistant Thermoplastic	THHN	194F	Dry locations
Moisture and Heat Resistant Thermoplastic	THW or THWN	167F	Dry and wet locations
Underground Feeder	UF	140F	Moisture resistant
Varnished Cambric	V	185F	Dry locations only

MOTOR SPEEDS		
NUMBER OF POLES	**SYNCHRONOUS SPEED**	**ACTUAL SPEED**
Two-pole Motor	7200 ÷ 2 = 3600 RPM	3450 RPM
Four-pole Motor	7200 ÷ 4 = 1800 RPM	1725 RPM
Six-pole Motor	7200 ÷ 6 = 1200 RPM	1150 RPM

TAP	COLOR
COMMON	WHITE
HIGH	BLACK
MEDIUM	YELLOW
LOW	RED
CAPACITOR	PURPLE (2 WIRES)

Table shows colors used by manufacturers to code wires used when making external connections to multi-speed motors.

HP	RUN WINDING	START WINDING
1/8	4.5 Ω	16 Ω
1/6	4.0 Ω	16 Ω
1/5	2.5 Ω	13 Ω
1/4	2.0 Ω	17 Ω

This table shows run and start winding resistance, in ohms, for common sizes of fractional horse-power single-phase

Coil No.	Pick Up Voltage	Drop Out Voltage	Continuous Voltage	Coil Ohms (approx.)
1	139-153	15-55	130	760
2	140-153	20-45	170	1400
3	159-172	35-77	256	3320
4	261-290	50-100	336	5180
5	280-310	50-100	395	7150
6	299-327	50-100	420	10000
7	323-352	60-135	495	11950

This table shows the pickup voltage ranges and other information on different relay coils that are available.

TO FIND	DIRECT CURRENT	ALTERNATING CURRENT AC	
	DC	SINGLE PHASE	THREE PHASE
Amperes when horsepower is known	$\dfrac{Hp \times 746}{E \times \%Eff.}$	$\dfrac{Hp \times 746}{E \times \%Eff \times PF.}$	$\dfrac{Hp \times 746}{1.73 \times E \times \%Eff \times PF.}$
Amperes when kilowatts are known	$\dfrac{kW \times 1000}{E}$	$\dfrac{kW \times 1000}{E \times PF}$	$\dfrac{kW \times 1000}{1.73 \times E \times PF}$
Amperes when KVA is known		$\dfrac{KVA \times 1000}{E}$	$\dfrac{KVA \times 1000}{1.73 \times E}$
Kilowatts	$\dfrac{I \times E}{1000}$	$\dfrac{I \times E \times PF}{1000}$	$\dfrac{I \times E \times 1.73 \times PF}{1000}$
KVA		$\dfrac{I \times E}{1000}$	$\dfrac{I \times E \times 1.73}{1000}$
Horsepower	$\dfrac{I \times E \times \%Eff.}{746}$	$\dfrac{I \times E \times \%Eff. \times PF}{746}$	$\dfrac{I \times E \times 1.73 \times \%Eff \times PF}{746}$

DIMENSIONS, TYPICAL RESISTANCES AND AMPACITY OF COMMERCIAL WIRE

GAUGE NO. (AWG)	DIAMETER BARE WIRE (INCHES)	OHMS PER 1000 FT. 70°F	OHMS PER 1000 FT. 167°F	CURRENT CAPACITY (AMPERES) COPPER TW UF	CURRENT CAPACITY (AMPERES) COPPER RH, RHW, THHW, THW THWN	CURRENT CAPACITY (AMPERES) ALUMINUM TW UF	CURRENT CAPACITY (AMPERES) ALUMINUM RH, RHW THHW,THW THWN
0000 (4/0)	0.460	0.050	.060	195	230	150	180
000 (3/0)	0.410	0.062	0.075	165	200	130	155
00 (2/0)	0.365	0.800	0.095	145	175	115	135
0 (1/0)	0.325	0.100	0.119	125	150	100	120
1	0.289	0.127	0.150	110	130	85	100
2	0.258	0.159	0.190	95	115	75	90
3	0.229	0.202	0.240	85	100	65	75
4	0.204	0.254	0.302	70	85	55	65
6	0.162	0.40	0.480	55	65	40	50
8	0.128	0.645	0.764	40	50	30	40
10	0.102	1.02	1.216	30*	30*	20	25
12	0.081	1.62	1.931	20*	20*	16	18
14	0.064	2.57	3.071	15*	15*		
16	0.051	4.10	4.884	10*	10*		
18	0.040	6.51	7.765	5*	5*		

MATERIAL SPECIFICATIONS AND PROPERTIES

NATIONAL COARSE THREADS		
TAP SIZE	THREADS PER INCH	DRILL SIZE
#5	40	#38
#6	32	#36
#8	32	#29
#10	24	#25
#12	24	#16
1/4"	20	#7
5/16"	18	F
3/8"	16	5/16"
7/16"	14	U
1/2"	13	27/64"
9/16"	12	31/64"
5/8"	11	17/32"
3/4"	10	21/32"
7/8"	9	49/64"
1"	8	7/8"

NATIONAL FINE THREADS		
TAP SIZE	THREADS PER INCH	DRILL SIZE
#5	44	#37
#6	40	#33
#8	36	#29
#10	32	#21
#12	28	#14
1/4"	28	#3
5/16"	24	I
3/8"	24	Q
7/16"	20	25/64"
1/2"	20	29/64"
9/16"	18	33/64"
5/8"	18	37/64"
3/4"	16	11/16"
7/8"	14	13/16"
1"	14	15/16"

NATIONAL PIPE THREADS

TAP SIZE	THREADS PER INCH	DRILL SIZE
1/8"	27	11/32"
1/4"	18	7/16"
3/8"	18	19/32"
1/2"	14	23/32"
3/4"	14	15/16"
1"	11-1/2	1-5/32"
1-1/4"	11-1/2	1-1/2"
1-1/2"	11-1/2	1-23/32"
2"	11-1/2	2-3/16"

MACHINE SCREW SIZES AND THREADS

MACHINE SCREW NUMBER	DIAMETER (IN.)	THREADS PER INCH	
		NATIONAL FINE	NATIONAL COARSE
2	.086	64	56
3	.099	56	48
4	.112	48	40
5	.125	44	40
6	.138	40	32
8	.164	36	32
10	.190	32	24
12	.216	28	24
1/4	.250	28	20
5/16	.3125	24	18
3/8	.375	24	16

Soldering Alloys

Solder	Composition Percent				Temperature	
	(Sn) Tin	(Pb) Lead	(Sb) Antimony	(Ag) Silver	Melts	Flows
50/50	50	50			360	420
40/60	40	60			360	460
60/40	60	40			360	375
95/5	95		5		452	464
Silver Solder	96			4	430	430
Silver Solder	94			6	430	535

ALUMINUM

SOLDER	MELTING POINT
Zinc Base	700°F to 820°F
Zinc Cadmium	510°F to 750°F
Tin Zinc	550°F and higher
Tin Lead	450°F and higher

TIPS VS. TUBING SIZE

TIP NO.	SOFT SOLDERING COPPER TUBING SIZE	BRAZING COPPER TUBING SIZE
#3	1/4" to 1-1/2"	1/4" to 7/8"
#4	1" to 2"	5/8" to 1-1/8"
#5	1-1/2" to 3"	7/8" to 1-5/8"
#6	2" to 4"	1-1/8" to 2-1/8"

CYLINDER SIZES FOR MATCHED SYSTEMS		
ACETYLENE		**OXYGEN**
10 cu. ft. (MC)	EQUALS	20 cu. Ft. (AA or R-Oxy)
40 cu. ft. (B)	EQUALS	40 cu. ft. (A)
60 cu. ft. (#2)	EQUALS	60 cu. ft. (J)
60 cu. ft. (#2)	ALSO EQUALS	80 cu. ft. (JJ)

CUTTING TIP PRESSURE SETTINGS			
METAL THICKNESS	**TIP SIZE**	**OXYGEN PSIG**	**ACETYLENE PSIG**
1/8"	000	20-25	5
1/4"	00	20-25	5
3/8"	0	25-30	5
1/2"	0	30-35	5
3/4"	1	30-40	5
1	2	35-50	6

INSULATION VALUE OF MATERIALS				
Insulation Material	Thickness (inches)	k*	C*	R Value
Air Space:				
non-reflective	3/4		0.99	1.01
reflective	3/4		0.29	3.48
reflective foil, 2 reflective surfaces	1	0.72		1.39
Aluminum siding over sheathing			1.61	0.61
Architectural glass			10.00	0.10
Asbestos-cement board		4.00		0.25
	1/8		33.00	0.03
	1/4		16.50	0.06
Asphalt roll roofing	0.048		6.50	0.15
Asphalt shingle	0.048		2.27	0.44
Balsam wood	1	0.27		3.70
Brick, common	1	5.00		0.20
Brick, face	1	9.00		0.11
Built-up roofing	3/8		3.00	0.33
Carpet and fibrous pad			0.48	2.08
Carpet with foam rubber pad			0.81	1.23
Cedar shingle			1.11	0.90
Cellulose, loose fill, blown	1	0.31		3.25
Cellular board	1	0.35		2.86
Cellular glass	1	0.38		2.63
	2		0.17	5.90
Cellulosics	1	0.29		3.50
Celotex	1	0.33		3.03
Cement fiber slab	1	0.50		2.00
Cement mortar	1	5.00		0.20
Cinder block, hollow	8		0.58	1.72
	12		0.53	1.89
Cinder block, hollow,				
with 1/2 inch of plaster	8-1/2		0.35	2.85
with 1/2 inch of plaster	12-1/2		0.33	3.03

INSULATION VALUE OF MATERIALS				
Insulation Material	Thickness (inches)	k*	C*	R Value
Clay tile				
hollow, 1 cell deep	3		1.25	0.80
hollow, 1 cell deep	4		0.90	1.11
hollow, 2 cells deep	6		0.66	1.52
hollow, 2 cells deep	8		0.54	1.85
hollow, 2 cells deep	10		0.45	2.22
hollow, 3 cells deep	12		0.40	2.50
Concrete block				
hollow	8		0.90	1.11
hollow	12		0.78	1.28
hollow with:				
lightweight aggregate	8		0.50	2.00
1/2 inch of plaster	8-1/2		0.49	2.04
1/2 inch of plaster	12-1/2		0.45	2.22
Concrete, poured, sand and gravel aggregate	1	12.50		0.08
Concrete, slab	4		3.13	0.32
Concrete, wall	8		1.56	0.64
Cork board	1	0.30		3.33
Cork tile	1/8		3.60	0.28
Felt, vapor-permeable			16.70	0.06
Fiberboard sheathing	1/2		0.76	1.32
	25/32		0.49	2.06
	1	0.42		2.36
Fiberglass batt	1	0.30		3.30
	2		0.16	6.30
	3-1/2		0.091	11.00
	6		0.053	19.00
	8		0.040	25.30
Fiberglass, loose fill	1	0.91		1.10
Floor tile, vinyl, etc.			20.00	0.05
Glass fiber, organic bonded		0.25		4.00
Glass fiber board	1	0.25		4.00

INSULATION VALUE OF MATERIALS				
Insulation Material	Thickness (inches)	k*	C*	R Value
Glass wool	1	0.27		3.76
Ground surface			2.00	0.50
Gypsum board	3/8		3.10	0.32
	1/2		2.22	0.45
	5/8		1.78	0.56
Gypsum board on gypsum lath	1/2		3.12	0.32
Hardboard	1/4		5.56	0.18
high density		0.82		1.22
high density, std. Temper		1.00		1.00
medium density	1	0.73		1.37
Hardwood	1	1.10		0.91
Hardwood floor	3/4	1.47		0.68
	25/32		1.43	0.70
Linoleum or rubber tile			20.00	0.05
Mineral fiber,		-		
loose fill, blown in	1	0.31		3.25
rock or glass	1	0.38		2.60
rock or glass	3-3/4 to 5			11.00
rock or glass	6½ - 8-3/4		0.05	19.00
rock or glass	10¼-13-3/4		0.03	30.00
resin binder	1	0.29		3.45
loose fill, blown in	7½ to 10			22.00
with resin binder		0.29		3.45
Mineral fiberboard	1	0.29		3.45
(wet-felted) acoustical tile	1	0.36		2.78
(wet-felted) roof insulation	1	0.34		2.94
(wet-molded) acoustical tile	1	0.42		2.38

INSULATION VALUE OF MATERIALS				
Insulation Material	Thickness (inches)	k*	C*	R Value
Mineral wool, batt	1	0.31		3.25
	1	0.24		4.16
	3 to 4		0.091	11.00
	5 ½ to 6 ½		0.053	19.00
	6 to 7 ½		0.045	22.00
	9 to 10		0.033	30.00
	12 to 13		0.026	38.00
Particleboard	5/8		1.22	0.82
high density		1.18		0.85
low density		0.54		1.85
medium density	1	0.94		1.06
underlayment	5/8		1.22	0.82
Perlite, expanded	1	0.33		3.03
expanded, organic bonded		0.36		2.78
loose fill	1	0.37		2.70
Plaster	1	8.33		0.12
Plaster and metal lath	3/4		7.69	0.13
Plaster, cement, sand	3/8		13.3	0.08
	3/4		6.66	0.15
	1	5.00		0.20
Plaster, gypsum				
lightweight agg	1/2		3.12	0.32
lightweight agg	5/8		2.67	0.39
lightweight agg	3/4		2.13	0.47
perlite agg				0.67
sand	1/2		11.10	0.09
sand	5/8		9.10	0.11
sand	1			0.18
on metal lath	3/4		7.70	0.13
Plasterboard	3/8		3.10	0.32
	1/2		2.22	0.45

INSULATION VALUE OF MATERIALS				
Insulation Material	Thickness (inches)	k*	C*	R Value
Plywood	1/4		3.20	0.31
	3/8		2.13	0.47
	1/2		1.60	0.62
	5/8		1.29	0.77
	1	0.80		1.25
Plywood (Douglas Fir)		0.80		1.25
Plywood or wood panels	3/4		1.07	0.93
Polycarbonate sheet	1/8		1.06	0.94
	3/16		1.01	0.99
	1/4		0.96	1.04
	3/8		0.88	1.14
	1/2		0.81	1.23
Polyisocyanurate. cellular	1/2		0.278	3.60
	1		0.139	7.20
	2		0.069	14.40
Polyisocyanurate, smooth skin	1	0.14		7.20
Polystyrene	1	0.28		3.57
cut cell	1	0.25		4.00
expanded, molded beads	1	0.26		3.85
foamed in place	1	0.27		3.75
smooth skin (Styrofoam)	1	0.20		5.00
Polyurethane				
expanded	1	0.14		7.00
expanded board	1	0.16		6.25
expanded. aged	1	0.16		6.30
Redwood	1	0.57		1.75
Rock cork	1	0.33		3.05
Rock wool batt	1	0.27		3.70
Roofing, 1-ply membrane	0.048		2.00	0.50
Rubber, expanded, board	1	0.22		4.55

INSULATION VALUE OF MATERIALS				
Insulation Material	Thickness (inches)	k*	C*	R Value
Sawdust	1	0.41		2.44
Sawdust/shavings	1	0.45		2.20
Sheep's wool	1	0.34		2.96
Slate shingle	1/2		20.00	0.05
Softwood	1	0.80		1.25
Stone	1	12.50		0.08
Structural insulation board	1/2		0.76	1.32
Stucco	1	5.00		0.20
Terrazzo	1	12.50		0.08
Tile, hollow	4		1.00	1.00
Urea-formaldehyde	1	0.24		4.20
Urethane, foamed in place	1	0.16		6.30
Vapor-seal, 2-layers of mopped 15-lb felt			8.35	0.12
Vapor-seal, plastic film				**
Vegetable Fiber Board Sheathing				
regular density	1/2		0.76	1.32
regular density	25/32		0.49	2.06
intermediate density	1/2		0.82	1.22
nail-base	½		0.88	1.14
Shingle backer	3/8		1.06	0.94
	5/16		1.28	0.78
sound deadening board	½		0.74	1.35
tile lay-in panels		0.40		2.50
	½		0.80	1.25
	¾		0.53	1.89
laminated paperboard		0.50		2.00
homo. Board from repulped paper		0.50		2.00
Vermiculite	1	0.47		2.13
Vermiculite, loose fill	1	0.45		2.20

INSULATION VALUE OF MATERIALS

Insulation Material	Thickness (inches)	k*	C*	R Value
Wall, vertical exterior, 15 mph wind			5.88	0.17
Wall, vertical interior, still air			1.47	0.68
Wood				
bevel lap siding	1/2		1.23	0.81
bevel lap siding	3/4		0.95	1.05
drop siding	3/4		1.27	0.79
drop siding	1	1.27		0.79
fiber, soft wood	1	0.30		3.33
fiberboard	1	0.59		1.69
fiberboard, acoustical tile	1/2		0.80	1.25
fiberboard, acoustical tile	3/4		0.53	1.89
shingle			1.06	0.94
shingle siding			1.15	0.87
shingle with insulating backer board			0.84	0.71
shingle, dbl subfloor	3/4		1.06	1.19
vertical tongue & groove	3/4		1.00	1.00

k* (in units of Btu in / ft²hr °F) is heat conductivity over a thickness of 1 inch and C* is heat conductance (in units of Btu/ ft² hr °F) over the specified thickness. **R Value** is the most common number used to compare the insulating properties of various materials and is typically marked on the wrapper or container of the insulator. The **R Value** is effectively the materials resistance to heat-flow and is based on the **k* and C*** values. **R Values** are the reciprocals of **k*** (which is 1/k) or **C*** (which is 1/C) for a given material.

**minimal.

TYPES OF COPPER PIPE

TYPE   CHARACTERISTICS

L -----Standard tubing used for interior. above ground plumbing. Uses include heating. air conditioning. steam. Can be used with sweat. flare. or compression fittings. Available in hard or soft types.

K -----Thick walled flexible tubing. Required for underground installations. Typical uses include water service. plumbing. heating. steam. gas. oil. oxygen. and applications where a thick walled tubing is required. Can be used with sweat. flare. or compression fittings. Available in hard or soft types.

M ------Typically used in interior heating and pressure applications. Wall thickness is less than types K and L. Normally used with sweat fittings. Available in hard or soft types.

DWV-----Drain - Waste - Vent. recommended for above ground. no pressure applications only. Use sweat fittings only. Available in hard type only.

When measuring copper pipe. sweat fittings are listed by their inside diameter (ID). compression fittings are listed by outside diameter (OD). Hard type comes in 20 foot straight lengths. Soft type comes in 20 foot straight lengths or 60 foot coils. Use 50 / 50 solid core solder and a high quality flux.

K TYPE & L TYPE
COPPER PIPE & TUBING

Nom. Size Inches	Actual OD Inches	K Type		L Type	
		Wall Th. Inch	Weight Lb. / ft	Wall Th. Inches	Weight Lb. / ft
1/4	0.375	0.035	0.145	0.030	0.126
3/8	0.500	0.049	0.269	0.035	0.198
1/2	0.625	0.049	0.344	0.040	0.265
5/8	0.750	0.049	0.418	0.042	0.362
3/4	0.875	0.065	0.641	0.045	0.455
1	1.125	0.065	0.839	0.050	0.655
1-1/4	1.375	0.065	1.040	0.055	0.884
1-1/2	1.625	0.072	1.360	0.060	1.140
2	2.125	0.083	2.060	0.070	1.750
2-1/2	2.625	0.095	2.930	0.080	2.480
3	3.125	0.109	4.000	0.090	3.330
3-1/2	3.625	0.120	5.120	0.100	4.290
4	4.125	0.134	6.510	0.110	5.380
5	5.125	0.160	9.670	0.125	7.610
6	6.125	0.192	13.9	0.140	10.2
8	8.125	0.271	25.9	0.200	19.3
10	10.125	0.338	40.3	0.250	30.1
12	12.125	0.405	57.8	0.280	40.4

OD = outside diameter   Wall Th. = wall thickness

M TYPE & DWV TYPE
COPPER PIPE AND TUBING

Nom. Size Inches	Actual OD Inches	M Type		DWV Type	
		Wall Th. Inches	Weight Lb. / ft	Wall Th. Inches	Weight Lb. / ft
1-1/4	1.375	0.042	0.682	0.040	0.65
1-1/2	1.625	0.049	0.940	0.042	0.81
2	2.125	0.058	0.146	0.042	1.07
2-1/2	2.625	0.065	2.030	---	---
3	3.125	0.072	2.680	0.045	1.69
3-1/2	3.625	0.083	3.580	---	---
4	4.125	0.095	4.660	0.058	2.87
5	5.125	0.109	6.660	0.072	4.43
6	6.125	0.122	8.920	0.083	6.10
8	8.125	0.170	16.5	---	---
10	10.125	0.212	25.6	---	---
12	12.125	0.254	36.7	---	---

OD = outside diameter   Wall Th. = wall thickness

TYPES OF PLASTIC PIPE

Type   Characteristics

PVC------Polyvinyl Chloride, Type 1, Grade 1.
Strong, rigid pipe resistant to a variety of acids and
bases. Some solvents and chlorinated hydrocarbons may
damage the pipe. Maximum usable temperature is
160°F. PVC can be used with water, gas, and drainage
systems, but NOT with hot water systems.

ABS------Acrylonitrile Butadiene Styrene, Type 1.
Strong rigid pipe resistant to most acids and bases. Some
solvents and chlorinated hydrocarbons may damage the
pipe. Maximum usable temperature 160°F. Most common use is DWV.

CPCV-----Chlorinated polyvinyl chloride.
Similar to PVC but designed for water piping up to
180°F. Pressure rating is 100 psi.

PE -------Polyethylene.
Flexible pipe for water systems (i.e., sprinklers). Not for
hot water.

PB -------Polybutylene.
Flexible pipe for hot and cold water systems. Use compression or banded type joints only.

Polypropylene –
Low pressure lightweight material. Highly resistant to
acids and bases and most solvents. Temperatures up to
180°F.

PVDF -----Polyvinylidene fluoride.
Strong pipe, resistant to abrasion, acids, basses, and
most solvents. Temperatures up to 280°F.

PVC PLASTIC PIPE

Nom. size inches	Actual OD inches	PVC Schedule 40		PVC schedule 80	
		Wall Th. Inches	Weight lb. / ft	Wall Th. inches	Weight lb. / ft
1/4	0.540	---	---	0.119	0.10
1/2	0.840	0.109	00.16	0.147	0.21
3/4	1.050	0.113	0.22	0.154	0.28
1	1.315	0.133	0.32	0.179	0.40
1-1/4	1.660	0.140	0.43	0.191	0.57
1-1/2	1.900	0.145	0.52	0.200	0.69
2	2.375	0.154	0.70	0.218	0.95
2-1/2	2.875	0.203	1.10	0.276	1.45
3	3.500	0.216	1.44	0.300	1.94
4	4.500	0.237	2.05	0.337	2.83
6	6.625	0.280	3.61	0.432	5.41
8	8.625	0.322	5.45	0.500	8.22
10	10.750	0.365	7.91	0.593	12.28
12	12.750	0.406	10.35	0.687	17.10

CPVC PLASTIC PIPE

Nom. size inches	Actual OD inches	CPVC schedule 40		CPVC schedule 80	
		Wall Th. Inches	Weight lb. / ft	Wall Th. inches	Weight lb. / ft
1/4	0.540	---	---	0.119	0.12
1/2	0.840	0.109	0.16	0.147	0.21
3/4	1.050	0.113	0.25	0.154	0.33
1	1.315	0.133	0.38	0.179	0.49
1-1/4	1.660	0.140	0.51	0.191	0.67
1-1/2	1.990	0.145	0.61	0.200	0.81
2	2.375	0.154	0.82	0.218	1.09
2-1/2	2.875	0.203	1.29	0.276	1.65
3	3.500	0.216	1.69	0.300	2.21
4	4.500	0.237	2.33	0.337	3.23
6	6.625	0.280	4.10	0.432	6.17
8	8.625	---	---	0.500	9.06

PVDF & POLYPROPYLENE PLASTIC PIPE

Nom. size inches	Actual OD inches	PVDF Schedule 40		Polypropylene Schedule 80	
		Wall Th. Inches	Weight lb. / ft	Wall Th. inches	Weight lb. / ft
1/2	0.840	0.147	0.24	0.147	0.14
3/4	1.050	0.154	0.33	0.154	0.19
1	1.315	0.179	0.49	0.179	0.27
1-1/4	1.660	0.191	---	0.191	0.38
1-1/2	1.900	0.200	0.81	0.200	0.45
2	2.375	0.218	1.13	0.218	0.62

Pipe Schedule Number = $1000 \times \frac{\text{psi internal pressure}}{\text{psi allowable fiber stress}}$

STEEL PIPE

Nominal size and OD inches	Schedule Numbers a - b - c (1)	Wall Th. Inches	Inside diameter inches	Pipe Weight lb. / ft
1/8 0.405	...-...-10S	0.049	0.307	0.18
	40-Std-40S	0.068	0.269	0.24
	80-XS-80S	0.095	0.215	0.31
1/4 0.540	...-...-10S	0.065	0.410	0.33
	40-Std-40S	0.088	0.364	0.42
	80-XS-80S	0.119	0.302	0.53
3/8 0.675	...-...-5S	0.065	0.710	0.53
	...-...-10S	0.065	0.545	0.42
	40-Std-40S	0.091	0.493	0.56
	80-XS-80S	0.126	0.423	0.73
1/2 0.840	...-...-5S	0.065	0.710	0.53
	...-...-10S	0.083	0.674	0.67
	40-Std-40S	0.109	0.622	0.85
	80-XS-80S	0.147	0.546	1.08
	160-...-...	0.187	0.466	1.30
	...-XXS-...	0.294	0.252	1.71
3/4 1.050	...-...-5S	0.065	0.920	0.68
	...-...-10S	0.083	0.884	0.85
	40-Std-40S	0.113	0.824	1.13
	80-XS-80S	0.154	0.742	1.47
	160-...-...	0.218	0.614	1.93
	...-XXS-...	0.308	0.434	2.44
1 1.315	...-...-5S	0.065	1.185	0.86
	...-...-10S	0.109	1.097	1.40
	40-Std-40S	0.133	1.049	1.67
	80-XS-80S	0.179	0.957	2.17
	160-...-...	0.250	0.815	2.84
	...-XXS-...	0.358	0.599	3.65

STEEL PIPE

Nominal size and OD inches	Schedule Numbers a - b - c (1)	Wall Th. Inches	Inside diameter inches	Pipe Weight lb. / ft
1-1/4 1.660	...-...-5S	0.065	1.530	1.10
	...-...-10S	0.109	1.442	1.80
	40-Std-40S	0.140	1.380	2.27
	80-XS-80S	0.191	1.287	2.99
	160-...-...	0.250	1.160	3.76
	...-XXS-...	0.382	0.896	5.21
1-1/2 1.900	...-...-5S	0.065	1.770	1.27
	...-...-10S	0.109	1.682	2.08
	40-Std-40S	0.145	1.610	2.71
	80-XS-80S	0.200	1.500	3.63
	160-...-...	0.281	1.338	4.85
	...-XXS-...	0.400	1.100	6.40
	...-...-...	0.525	0.850	7.71
	...-...-...	0.650	0.600	8.67
2 2.375	...-...-5S	0.065	2.245	1.6-
	...-...-10S	0.109	2.157	2.63
	40-Std-40S	0.154	2.067	3.65
	80-XS-80S	0.218	1.939	5.02
	160-...-...	0.343	1.689	7.44
	...-XXS-...	0.436	1.503	9.02
	...-...-...	0.562	1.251	11
	...-...-...	0.687	1.001	12
2-1/2 2.875	...-...-5S	0.083	2.709	2.0
	...-...-10S	0.120	2.635	3.5
	40-Std-40S	0.203	2.469	5.8
	80-XS-80S	0.276	2.323	7.7
	160-...-...	0.375	2.125	10
	...-XXS-...	0.552	1.771	14
	...-...-...	0.675	1.525	16
	...-...-...	0.800	1.275	18

STEEL PIPE

Nominal size and OD inches	Schedule Numbers a - b - c (1)	Wall Th. Inches	Inside diameter inches	Pipe Weight lb. / ft
	...-...-5S	0.083	3.334	3.0
	...-...-10S	0.120	3.260	4.3
	40-Std-40S	0.216	3.068	7.6
3	80-XS-80S	0.300	2.900	10.2
3.500	160-...-...	0.437	2.626	14.3
	...-XXS-...	0.600	2.300	19
	...-...-...	0.725	2.050	21
	...-...-...	0.850	1.800	24
	...-...-5S	0.083	3.834	3.5
	...-...-10S	0.120	3.760	5.0
3-1/2	40-Std-40S	0.226	3.548	9.1
4.000	80-XS-80S	0.318	3.364	12
	...-XXS-...	0.636	2.728	23
	...-...-5S	0.083	4.334	3.9
	...-...-10S	0.120	4.260	5.6
	...-...-...	0.188	4.124	8.6
	40-Std-40S	0.237	4.026	11
4	80-XS-80S	0.337	3.826	15
4.500	120-...-...	0.437	3.626	19
	...-...-...	0.500	3.500	21
	160-...-...	0.531	3.438	23
	...-XXS-...	0.674	3.152	28
	...-...-...	0.800	2.900	32
	...-...-...	0.925	2.650	35

STEEL PIPE

Nominal size and OD inches	Schedule Numbers a - b - c (1)	Wall Th. Inches	Inside diameter inches	Pipe Weight lb. / ft
	...-...-5S	0.109	5.345	6.3
	...-...-10S	0.134	5.295	7.8
5	40-Std-40S	0.258	5.047	15
5.563	80-XS-80S	0.375	4.813	21
	120-...-...	0.500	4.563	27
	160-...-...	0.625	4.313	33
	...-XXS-...	0.750	4.063	38
	...-...-...	0.875	3.813	44
	...-...-...	1.000	3.563	48
	...-...-5S	0.109	6.407	5.4
	...-...-10S	0.134	6.357	9.3
	...-...-...	0.219	6.187	15
	40-Std-40S	0.280	6.065	19
6	80-XS-80S	0.432	5.761	28
6.625	120-...-...	0.562	5.501	36
	160-...-...	0.718	5.189	45
	...-XXS-...	0.864	4.897	53
	...-...-...	1.000	4.625	60
	...-...-...	1.125	4.375	66

STEEL PIPE

Nominal size and OD inches	Schedule Numbers a - b - c (1)	Wall Th. Inches	Inside diameter inches	Pipe Weight lb. / ft
	...-...-5S	0.109	8.407	9.9
	...-...-10S	0.148	8.329	13
	...-...-...	0.219	8.187	20
	20-...-...	0.250	8.125	22
8	30-...-...	0.277	8.071	25
8.625	40-Std-40S	0.322	7.981	29
	60-...-...	0.406	7.813	36
	80-XS-80S	0.500	7.625	43
	100-...-...	0.593	7.439	51
	120-...-...	0.718	7.189	61
	140-...-...	0.812	7.001	68
	160-...-...	0.906	6.813	75
	...-...-...	1.000	6.625	81
	...-...-...	1.125	6.375	90
	...-...-5S	0.134	10.482	15
	...-...-10S	0.165	10.420	19
	...-...-...	0.219	10.312	25
	20-...-...	0.250	10.250	28
	30-...-...	0.307	10.136	34
	40-Std-40S	0.365	10.020	40
10	60-...-...	0.500	9.750	55
10.750	80-XS-80S	0.593	9.564	64
	100-...-...	0.718	9.314	77
	120-...-...	0.843	9.064	89
	...-...-...	0.875	9.000	92
	140-...-...	1.000	8.750	104
	160-...-...	1.125	8.500	116
	...-...-...	1.250	8.250	127
	...-...-...	1.500	7.750	148

STEEL PIPE

Nominal size and OD inches	Schedule Numbers a - b - c (1)	Wall Th. Inches	Inside diameter inches	Pipe Weight lb. / ft
	...-...-5S	0.156	12.438	21
	...-...-10S	0.180	12.390	24
	20-...-...	0.250	12.250	33
	30-...-...	0.330	12.090	44
	...-Std-40S	0.375	12.000	50
	40-...-...	0.406	11.938	54
12	...-XS-80S	0.500	11.750	65
12.750	60-...-...	0.562	11.626	73
	80-...-...	0.687	11.376	89
	...-...-...	0.750	11.250	96
	100-...-...	0.843	11.064	107
	...-...-...	0.875	11.000	111
	120-...-...	1.000	10.750	125
	140-...-...	1.125	10.500	140
	...-...-...	1.250	10.250	154
	160-...-...	1.312	10.126	160

STEEL PIPE

Nominal size and OD inches	Schedule Numbers a - b - c (1)	Wall Th. Inches	Inside diameter inches	Pipe Weight lb. / ft
	...-...-5S	0.156	13.688	23
	...-...-10S	0.188	13.624	28
	...-...-...	0.210	13.580	31
	...-...-...	0.219	13.562	32
	10-...-...	0.250	13.500	37
	...-...-...	0.281	13.438	41
	20-...-...	0.312	13.376	46
	...-...-...	0.344	13.312	50
14	30-Std-...	0.375	13.250	55
14.000	40-...-...	0.437	13.126	63
	...-...-...	0.469	13.062	68
	...-XS-...	0.500	13.000	72
	60-...-...	0.593	12.814	85
	...-...-...	0.625	12.750	89
	80-...-...	0.750	12.500	106
	100-...-...	0.937	12.126	131
	120-...-...	1.093	11.814	151
	140-...-...	1.250	11.500	170
	160-...-...	1.406	11.188	189
	...-...-5S	0.165	15.670	28
	...-...-10S	0.188	15.624	32
	10-...-...	0.250	15.500	42
	20-...-...	0.312	15.376	52
16	30-Std-...	0.375	15.250	63
16.000	40-XS-...	0.500	15.000	83
	60-...-...	0.656	14.688	107
	80-...-...	0.843	14.314	136
	100-...-...	1.031	13.938	165
	120-...-...	1.218	13.564	192
	140-...-...	1.437	13.126	224
	160-...-...	1.593	12.814	245

STEEL PIPE

Nominal size and OD inches	Schedule Numbers a - b - c (1)	Wall Th. Inches	Inside diameter inches	Pipe Weight lb. / ft
	...-...-5S	0.165	17.670	31
	...-...-10S	0.188	17.624	36
	10-...-...	0.250	17.500	47
	20-...-...	0.312	17.376	59
	...-Std-...	0.375	17.250	71
18	30-...-...	0.437	17.126	82
18.000	...-XS-...	0.500	17.000	93
	40-...-...	0.562	16.876	105
	60-...-...	0.750	16.500	138
	80-...-...	0.937	16.126	171
	100-...-...	1.156	15.688	208
	120-...-...	1.375	15.250	244
	140-...-...	1.562	14.876	274
	160-...-...	1.781	14.438	308
	...-...-5S	0.188	19.634	40
	...-...-10S	0.218	19.564	46
	10-...-...	0.250	19.500	53
	20-Std-...	0.375	19.250	79
	30-XS-...	0.500	19.000	104
	40-...-...	0.593	18.814	123
20	60-...-...	0.812	18.376	166
20.000	...-...-...	0.875	18.250	179
	80-...-...	1.031	17.938	209
	100-...-...	1.281	17.438	256
	120-...-...	1.500	17.000	296
	140-...-...	1.750	16.500	341
	160-...-...	1.968	16.064	379

(1) a = ANSI B36.10 Steel Pipe Schedule numbers
    b = ANSI B36.10 Steel Pipe nominal wall thickness
    c = ANSI B36.19 Stainless Steel Schedule numbers
Std = standard   XS = Extra strong   XXS = Double extra strong

DRILL & CUTTING LUBRICANTS

Material	Machine Process		
	Drilling	Threading	Lathe
Aluminum	Soluble oil Kerosene	Soluble oil Kerosene	Soluble oil
Brass	Dry Soluble oil Kerosene	Soluble oil Lard Oil	Soluble oil
Bronze	Dry Soluble oil Mineral oil	Soluble oil Lard oil	Soluble oil
Cast Iron	Dry Air jet	Dry Sulphurized oil	Dry Soluble oil
Copper	Dry Soluble oil Mineral lard oil	Soluble oil Lard oil	Soluble oil
Malleable Iron	Dry	Lard oil	Soluble oil
Monel metal	Soluble oil	Lard oil	Soluble oil
Steel alloys	Soluble oil Sulphurized oil	Sulphurized oil Lard oil	Soluble oil
Steel, machine	Soluble oil Sulphurized oil Lard oil	Soluble oil Mineral lard oil	Soluble oil
Steel, tool	Soluble oil Sulphurized oil Mineral lard oil	Sulphurized oil Lard oil	Soluble oil

DRILLING SPEEDS vs MATERIAL		
Material	Speed rpm	Description
Cast Iron	6000 to 6500	1/16 inch drill
	3500 to 4500	1/8 inch drill
	2500 to 3000	3/16 inch drill
	2000 to 2500	1/4 inch drill
	1500 to 2000	5/16 inch drill
	1500 to 2000	3/8 inch drill
	1000 to 1500	>7/16 inch drill
Glass	700	Special metal tube
Plastics	6000 to 6500	1/16 inch drill
	5000 to 6000	1/8 inch drill
	3500 to 4000	3/16 inch drill
	3000 to 3500	1/4 inch drill
	2000 to 2500	5/16 inch drill
	1500 to 2000	3/8 inch drill
	500 to 1000	>7/16 inch drill
Soft Metals (copper)	6000 to 6500	1/16 inch drill
	6000 to 6500	1/8 inch drill
	5000 to 6000	3/16 inch drill
	4500 to 5000	1/4 inch drill
	3500 to 4000	5/16 inch drill
	3000 to 3500	3/8 inch drill
	1500 to 2500	>7/16 inch drill
Steel	5000 to 6500	1/16 inch drill
	3000 to 4000	1/8 inch drill
	2000 to 2500	3/16 inch drill
	1500 to 2000	1/4 inch drill
	1000 to 1500	5/16 inch drill
	1000 to 1500	3/8 inch drill
	500 to 1000	>7/16 inch drill
Wood	4000 to 6000	Carving and routing
	3800 to 4000	All woods, 0 to 1/4 inch drills
	3100 to 3800	All woods, 1/4 to 1/2 inch drills
	2300 to 3100	All woods, 1/2 to 3/4 inch drills
	2000 to 2300	All woods, 3/4 to 1 inch drills
	700 to 2000	All woods, 1 inch drills, fly cutters
	<700	and multi-spur bits

STEEL SHEET GAUGES

Gauge Nbr.	Steel Weight lbs. Per sq. foot	Thickness Inches U.S. Std. Gauge	Manufac- turers Standard	Weight lbs./sq. ft. Galvan- ized Sheet	Stain- less Steel
7/0	20.00	0.5000	...	...	...
6/0	18.75	0.4687	...	...	...
5/0	17.50	0.4375	...	...	...
4/0	16.25	0.4062	...	...	...
3/0	15.00	0.3750	...	...	...
2/0	13.75	0.3437	...	...	...
0	12.50	0.3125	...	...	...
1	11.25	0.2812	...	...	...
2	10.62	0.2656	...	...	...
3	10.00	0.2500	0.2391	...	...
4	9.37	0.2344	0.2242	...	...
5	8.75	0.2187	0.2092	...	...
6	8.12	0.2031	0.1943	...	...
7	7.50	0.1875	0.1793	...	...
8	6.87	0.1719	0.1644	...	...
9	6.25	0.1562	0.1495	...	...
10	5.62	0.1406	0.1345	5.7812	5.7937
11	5.00	0.1250	0.1196	5.1562	5.1500
12	4.37	0.1094	0.1046	4.5312	4.5063
13	3.75	0.0937	0.0897	3.9062	3.8625
14	3.12	0.0781	0.0747	3.2812	3.2187
15	2.81	0.0703	0.0673	2.9687	2.8968
16	2.50	0.0625	0.0598	2.6562	2.5750
17	2.25	0.0562	0.0538	2.4062	2.3175
18	2.00	0.0500	0.0478	2.1562	2.0600

STEEL SHEET GAUGES

| Gauge Nbr. | Steel Weight lbs. Per sq. foot | Thickness Inches | | Weight lbs./sq. ft. | |
		U.S. Std. Gauge	Manufac- turers Standard	Galvan- ized Sheet	Stain- less Steel
19	1.75	0.0437	0.418	1.9062	1.8025
20	1.50	0.0375	0.0359	1.6562	1.5450
21	1.37	0.0344	0.0329	1.5312	1.4160
22	1.25	0.0312	0.0299	1.4062	1.2875
23	1.12	0.0281	0.0269	1.2812	1.1587
24	1.00	0.0250	0.0239	1.1562	1.0300
25	0.875	0.0219	0.0209	1.0312	0.9013
26	0.750	0.0187	0.0179	0.9062	0.7725
27	0.687	0.0172	0.0164	0.8437	0.7081
28	0.625	0.0156	0.0149	0.7812	0.6438
29	0.562	0.0141	0.0135	0.7187	0.5794
30	0.500	0.0125	0.0120	0.6562	0.5150
31	0.437	0.0109	0.0105	...	...
32	0.406	0.0102	0.0097	...	...
33	0.375	0.0094	0.0090	...	...
34	0.344	0.0086	0.0082	...	...
35	0.312	0.0078	0.0075	...	...
36	0.281	0.0070	0.0067	...	...
37	0.266	0.0066	0.0064	...	...
38	0.250	0.0062	0.0060	...	...
39	0.234	0.0059	...	...	...
40	0.219	0.0055	...	...	...
41	0.211	0.0053	...	...	...
42	0.203	0.0051	...	...	...
43	0.195	0.0049	...	...	...
44	0.187	0.0047	...	...	...

STEEL PLATE SIZES

Thickness inches	Weight lbs./sq. foot	Thickness inches	Weight lbs./sq. foot
3/16	7.65	2-1/8	86.70
1/4	10.20	2-1/4	91.80
5/16	12.75	2-1/2	102.00
3/8	15.30	2-3/4	112.20
7/16	17.85	3	122.40
1/2	20.40	3-1/4	132.60
9/16	22.95	3-1/2	142.80
5/8	25.50	3-3/4	153.00
11/16	28.05	4	163.20
3/4	30.60	4-1/4	173.40
13/16	33.15	4-1/2	183.60
7/8	35.70	5	204.00
1	40.80	5-1/2	224.40
1-1/8	45.90	6	244.80
1-1/4	51.00	6-1/2	265.20
1-3/8	56.10	7	285.60
1-1/2	61.20	7-1/2	306.00
1-5/8	66.30	8	326.40
1-3/4	71.40	9	367.20
1-7/8	76.50	10	408.00
2	81.60		

CHANNEL STEEL SPECS

Size (Bar) Inches	Weight lbs./foot	Structural Channel Size Inches	Weight lbs./foot
3/4 x 5/16 x 1/8............	0.50	C 5x1-3/4x0.190...........	6.7
3/4 x 3/8 x 1/8............	0.56	x1-7/8x0.325...........	9.0
7/8 x 3/8 x 1/8............	0.61	C 6x1-7/8x0.200...........	8.2
7/8 x 7/16 x 1/8............	0.69	x2x0.314................	10.5
1 x 3/8 x 1/8............	0.68	x2-1/8x0.437............	13.0
1 x 1/2 x 1/8............	0.84	MC 6x2-1/2x0.310........	12.0
1-1/8 x 9/16 x 3/16........	1.16	MC 6x3x0.316...........	15.1
1-1/4 x 1/2 x 1/8............	1.01	x3x0.375..............	16.3
1-1/2 x 1/2 x 1/8............	1.12	MC 6x3-1/2x0.340........	15.3
1-1/2 x 9/16 x 3/16........	1.44	x3-1/2x0.379............	18.0
1-1/2 x 3/4 x 1/8............	1.17	C 7x2-1/8x0.210...........	9.8
1-1/2 x 1-1/2 x 3/16........	2.65	x2-1/4x0.314............	12.25
1-3/4 x 1/2 x 3/16........	1.55	x2-1/4x0.419............	14.75
2 x 1/2 x 1/8............	1.43	MC 7x3x0.375...........	17.6
2 x 9/16 x 3/16........	1.86	x3-1/2x0.352............	19.1
2 x 5/8 x 1/4............	2.28	x3-5/8x0.503............	22.7
2 x 1 x 1/8............	1.59	MC 8x1-7/8x0.179........	8.50
2 x 1 x 3/16............	2.32	C 8x2-1/4x0.220...........	11.5
2-1/2 x 5/8 x 3/16........	2.27	x2-3/8x0.303............	13.75

Structural Channel			x2-1/2x0.487............	18.75
C = Standard Channel			MC 8x3x0.353...........	18.7
MC = Misc. Channel			x3x0.400..............	20.0
Size Inches	Weight lbs./foot		x3-1/2x0.377............	21.4
C 3x1-3/8x0.170............	4.1		x3-1/2x0.427............	22.8
x1-1/2x0.258............	5.0		C 9x2-3/8x0.233...........	13.4
x1-5/8x0.356............	6.0		x2-1/2x0.285............	15.0
MC 3x1-7/8x0.312............	7.1		x2-5/8x0.448............	20.0
x1-7/8x0.500............	9.0		MC 9x3-1/2x0.400........	23.9
C 4x1-5/8x0.184............	5.4		x3-1/2x0.450............	25.4
x1-5/8x0.247............	6.25		MC 10x1-1/8x0.152........	6.5
x1-3/4x0.321............	7.25		x1-1/2x0.170........	8.4
MC 4x2-1/2x0.500...........	13.8		x3-3/8x0.290............	22.0
			x3-3/8x0.380............	25.0

CHANNEL & ANGLE STEEL SPECS

Structural Channel		ANGLE STEEL	
Size Inches	**Weight lbs./foot**	**Size Inches**	**Weight lbs./foot**
C 10x2-5/8x0.240..........	15.3	1/2 x 1/2 x 1/8..............	0.38
x2-3/4x0.379........	20.0	5/8 x 5/8 x 1/8..............	0.48
x2-7/8x0.526........	25.0	3/4 x 3/4 x 1/8..............	0.59
x3x0.673..............	30.0	x 3/32..............	0.463
MC 10x4x0.425..........	28.5	x 3/16..............	0.84
x4-1/8x0.575........	33.6	7/8 x7/8 x 1/8..............	0.70
x4-3/8x0.796........	41.1	1 x 5/8 1/8..............	0.64
C 12x3x0.282..........	20.7	1 x 3/4 x 1/8..............	0.70
x3x0.387..............	25.0	1 x 1 x 1/8..............	0.80
x3-1/8x0.510........	30.0	x 3/16..............	1.16
MC 12x1-1/2x0.190......	10.6	x 1/4..............	1.49
x3-5/8x0.370........	31.0	1-1/8 x 1-1/8 x 1/8..............	0.90
x3-3/4x0.467........	35.0	1-1/4 x 1-1/4 x 1/8..............	1.01
x3-7/8x0.590........	40.0	x 3/16..............	1.48
x4x0.712..............	45.0	x 1/4..............	1.92
x4-1/8x0.835........	50.0	1-3/8 x 7/8 x 1/8..............	0.91
MC 13x4x0.375..............	31.8	x 3/16..............	1.32
x4-1/8x0.447........	35.0	1-1/2 x 1-1/4 x 3/16..............	1.64
x4-1/8x0.560........	40.0	1-1/2 x 1-1/2 x 1/8..............	1.23
x4-3/8x0.787........	50.0	x 3/16..............	1.80
C 15x3-3/8x0.400..........	33.9	x 1/4..............	2.34
x3-1/2x0.520........	40.0	x 5/16..............	2.86
x3-3/4x0.716.........	50.0	x 3/8..............	3.35
MC 18x4x0.450..............	42.7	1-3/4 x 1-1/4 x 1/8..............	1.23
x4x0.500..............	45.8	x 1/4..............	2.34
x4-1/8x0.600........	51.9	1-3/4 x 1-3/4 x 1/8..............	1.44
x4-1/8x0.700........	58.0	x 3/16..............	2.12
		x 1/4..............	2.77
		x 5/16..............	3.39
		x 3/8..............	3.99
		2 x 1-1/4 x 3/16..............	1.96
		x 1/4..............	2.55

ANGLE STEEL SPECS

Angle Steel		Angle Steel	
Size Inches	Weight lbs./foot	Size Inches	Weight lbs./foot
2 x 1-1/2 x 1/8............	1.44	3 x 2 x 3/8............	6.6
x 3/16............	2.12	x 1/2............	8.5
x 1/4............	2.77	3 x 3 x 3/16............	3.7
2 x 2 x 1/8............	1.65	x 1/4............	4.9
x 3/16............	2.44	x 5/16............	6.1
x 1/4............	3.19	x 3/8............	7.2
x 5/16............	3.92	x 7/16............	8.3
x 3/8............	4.70	x 1/2............	9.4
x 1/2............	6.00	3-1/2 x 2-1/2 x 1/4............	4.9
2-1/4 x 1-1/2 x 3/16............	2.28	x 5/16............	6.1
x 1/4............	2.98	x 3/8............	7.2
2-1/4 x 2-1/4 x 3/16............	2.75	x 1/2............	9.4
x 1/4............	3.62	3-1/2 x 3 x 1/4............	5.4
x 5/16............	4.50	x 5/16............	6.6
x 3/8............	5.30	x 3/8............	7.9
2-1/2 x 1-1/2 x 3/16............	2.44	x 1/2............	10.2
x 1/4............	3.19	3-1/2 x 3-1/2 x 1/4............	5.8
x 5/16............	3.92	x 5/16............	7.2
2-1/2 x 2 x 1/8............	1.86	x 3/8............	8.5
x 3/16............	2.75	x 7/16............	9.8
x 1/4............	3.62	x 1/2............	11.1
x 5/16............	4.50	4 x 3 x 1/4............	5.8
x 3/8............	5.30	x 5/16............	7.2
x 1/2............	6.74	x 3/8............	8.5
2-1/2 x 2-1/2 x 3/16............	3.07	x 7/16............	9.8
x 1/4............	4.10	x 1/2............	11.1
x 5/16............	5.00	x 5/8............	13.6
x 3/8............	5.90	4 x 3-1/2 x 1/4............	6.2
x 1/2............	7.70	x 5/16............	7.7
3 x 2 x 3/16............	3.07	x 3/8............	9.1
x 1/4............	4.1	x 7/16............	10.6
x 5/16............	5.0	x 1/2............	11.9

ANGLE STEEL SPECS

Angle Steel		Angle Steel	
Size Inches	Weight lbs./foot	Size Inches	Weight lbs./foot
4 x 4 x 1/4	6.6	6 x 4 x 1/2	16.2
x 5/16	8.2	x 5/8	20.0
x 3/8	9.8	x 3/4	23.6
x 7/16	11.3	x 7/8	27.2
x 1/2	12.8	x 9/16	18.1
x 5/8	15.7	6 x 6 x 5/16	12.5
x 3/4	18.5	x 3/8	14.9
5 x 3 x 1/4	6.6	x 7/16	17.2
x 5/16	8.2	x 1/2	19.6
x 3/8	9.8	x 9/16	21.9
x 7/16	11.3	x 5/8	24.2
x 1/2	12.8	x 3/4	28.7
5 x 3-1/2 x 1/4	7.0	x 7/8	33.1
x 5/16	8.7	x 1	37.4
x 3/8	10.4	7 x 4 x 3/8	13.6
x 7/16	12.0	x 7/16	15.8
x 1/2	13.6	x 1/2	17.9
x 5/8	16.8	x 5/8	22.1
x 3/4	19.8	x 3/4	26.2
5 x 5 x 5/16	10.3	8 x 4 x 7/16	17.2
x 3/8	12.3	x 1/2	19.6
x 7/16	14.3	x 9/16	21.9
x 1/2	16.2	x 5/8	24.2
x 5/8	20.0	x 3/4	28.7
x 3/4	23.6	x 7/8	33.1
x 7/8	27.2	x 1	37.4
6 x 3-1/2 x 5/16	9.8	8 x 6 x 7/16	20.4
x 3/8	11.7	x 1/2	23.0
x 1/2	15.3	x 9/16	25.7
6 x 4 x 5/16	10.3	x 5/8	28.5
x 3/8	12.3		
x 7/16	14.3		

GENERAL
SCIENCE

BOILING AND FREEZING TEMPERATURES

BOILING TEMPERATURE		FREEZING TEMPERATURE	
WATER	212	WATER	32
ETHYL ALCOHOL	173	FRUIT & VEG.	30
CHLOROFORM	143	SEAFOOD	28
BUTANE	31	BEEF AND PORK	28
AMMONIA	-28	POULTRY	27
PROPANE	-43	CARBON TET.	-9
CARBON DIOXIDE	-109	LINSEED OIL	-11
ACETYLENE	-118	CHLOROFORM	-81
OXYGEN	-287	AMMONIA	-107
NITROGEN	-320	ACETONE	-139
HYDROGEN	-423	ETHER	-177
HELIUM	-452	ETHYL ALCOHOL	-179

SPECIFIC HEATS

SUBSTANCE	SPECIFIC HEAT (LB.)	SUBSTANCE	SPECIFIC HEAT (LB.)
ACETONE	.514	CHICKEN	3.316
ALCOHOL	.680	CHLOROFORM	2.340
AMMONIA	1.090	COPPER	.095
BACON	1.474	FISH	3.550
BEEF	2.345	ICE	.487
BEER	3.852	IRON	1.373
BENZINE	.412	ORANGES	3.751
BREAD	1.993	PEACHES	3.818
BUTTER	1.373	POPCORN	1.172
CHEESE	2.077	WATER	1.000

HEATING VALUES	
FUEL	HEAT RELEASED Btu / lb.
COAL	
Bituminous	12.000 to 15.000
Anthracite	13.000 to 14.000
GAS	
Natural	1000 to 1100*
Manufactured	500 to 600
LP (liquefied petroleum)	2500 to 3200
* check with local gas company	
FUEL OIL	Btu / gal.
Grade 1	137.000
Grade 2	140.000
Grade 3*	140.000
Grade 4*	141.000
Grade 5**	148.000
Grade 6**	152.000
* not in common use	
** No. 5 & 6 are high viscosity and require preheating	

ABSOLUTE PRESSURE VALUES				COMPOUND GAUGE READING in. Hg psig	H$_2$O (BOILING POINT) °F
psia	in.Hg	mm Hg	microns		
14.696	29.291	759.999	759.999	00.000	212.00
14.000	28.504	724.007	724.007	1.418	209.56
13.000	26.468	672.292	672.292	3.454	205.88
12.000	24.432	620.577	620.577	5.490	201.96
11.000	22.396	568.862	568.862	7.526	197.75
10.000	20.360	517.147	517.147	9.617	193.21
9.000	18.324	465.432	436.432	11.598	188.28
8.000	16.288	413.718	413.718	13.634	182.86
7.000	14.252	362.003	362.003	15.670	176.85
6.000	12.216	310.289	310.289	17.706	170.06
5.000	10.180	285.573	258.573	19.742	162.24
4.000	8.144	206.859	206.859	21.778	152.97
3.000	6.108	155.144	155.144	23.813	141.48
2.000	4.072	103.430	103.430	25.849	126.08

130

ABSOLUTE PRESSURE VALUES				COMPOUND GAUGE READING in. Hg psig	H_2O (BOILING POINT) °F
psia	in.Hg	mm Hg	microns		
1.000	2.036	51.715	51.715	27.885	101.74
0.900	1.832	46.543	46.543	28.089	98.24
0.800	1.629	41.371	41.371	28.292	94.38
0.700	1.425	36.200	36.200	28.496	90.08
0.600	1.222	31.029	31.029	28.699	85.21
0.500	1.180	25.857	25.857	28.903	79.58
0.400	0.814	20.686	20.686	29.107	72.86
0.300	0.611	15.514	15.514	29.310	64.47
0.200	0.407	10.343	10.343	29.514	53.14
0.100	0.204	5.171	5.171	29.717	35.00
0.000	0.000	0.000	0.000	29.921	-

NOTE: psia X 2.035966 = in. Hg   psia X 51.715 = mm Hg   psia X 51.715 = microns

CONVERSION TABLES
AND
EQUIVALENTS

FAHRENHEIT TO CELSIUS (CENTIGRADE)	
Degrees Fahrenheit	Degrees Celsius
-20	-28.8
-15	-26.1
-10	-23.3
-5	-20.6
0	-17.8
1	-17.2
2	-16.7
3	-16.1
4	-15.6
5	-15.0
10	-12.2
15	-9.4
20	-6.7
25	-3.9
30	-1.1
35	1.7
40	4.4
45	7.2
50	10.0
55	12.8
60	15.6
65	18.3
70	21.1
75	23.9
80	26.7
85	29.4
90	32.2
95	35.0
100	37.8
105	40.6
110	43.3
115	46.1

FAHRENHEIT TO CELSIUS (CENTIGRADE)	
Degrees Fahrenheit	**Degrees Celsius**
120	48.9
125	51.7
130	54.4
135	57.2
140	60.0
145	62.8
150	65.6
155	68.3
160	71.1
165	73.9
170	76.7
175	79.4
180	82.2
185	85.0
190	87.8
195	90.6
200	93.8
205	96.1
210	98.9
212	100.0
215	101.7
220	104.4
225	107.2
230	110.0
235	112.8
240	115.6
245	118.3
250	121.1
255	123.9
260	126.6
265	129.4

DECIMAL EQUIVALENTS

Fraction Inch	Decimal Inch	Decimal Millimeter
1/64	0.0156	0.3969
1/32	0.0313	0.7937
3/64	0.0469	1.1906
1/16	0.0625	1.5875
5/64	0.0781	1.9844
3/32	0.0938	2.3812
7/64	0.1094	2.7781
1/8	0.1250	3.1750
9/64	0.1406	3.5719
5/32	0.1563	3.9687
11/64	0.1719	4.3656
3/16	0.1875	4.7625
13/64	0.2031	5.1594
7/32	0.2188	5.5562
15/64	0.2344	5.9531
1/4	0.2500	6.3500
17/64	0.2656	6.7469
9/32	0.2813	7.1437
19/64	0.2969	7.5406
5/16	0.3125	7.9375
21/64	0.3281	8.3344
11/32	0.3438	8.7312
23/64	0.3594	9.1281
3/8	0.3750	9.5250
25/64	0.3906	9.9219
13/32	0.4063	10.3187
27/64	0.4219	10.7156
7/16	0.4375	11.1125
29/64	0.4531	11.5094
15/32	0.4688	11.9062
31/64	0.4844	12.3031
1/2	0.5000	12.7000

DECIMAL EQUIVALENTS

Fraction Inch	Decimal Inch	Decimal Millimeter
33/64	0.5156	13.0969
17/32	0.5312	13.4937
35/64	0.5469	13.8906
9/16	0.5625	14.2875
37/64	0.5781	14.6844
19/32	0.5937	15.0812
39/64	0.6094	15.4781
5/8	0.6250	15.8750
41/64	0.6406	16.2719
21/32	0.6562	16.6687
43/64	0.6719	17.0656
11/16	0.6875	17.4625
45/64	0.7031	17.8594
23/32	0.7188	18.2562
47/64	0.7344	18.6531
3/4	0.7500	19.0500
49/64	0.7656	19.4469
25/32	0.7812	19.8437
51/64	0.7969	20.2406
13/16	0.8125	20.6375
53/64	0.8281	21.0344
27/32	0.8437	21.4312
55/64	0.8594	21.8281
7/8	0.8750	22.2250
57/64	0.8906	22.6219
29/32	0.9062	23.0187
59/64	0.9219	23.4156
15/16	0.9375	23.8125
61/64	0.9531	24.2094
31/32	0.9687	24.6062
63/64	0.9844	25.0031
1	1.0000	25.4016

POUNDS TO KILOGRAMS

lb.	0 kg	1 kg	2 kg	3 kg	4 kg	5 kg	6 kg	7 kg	8 kg	9 kg
-		0.454	0.907	1.361	1.814	2.268	2.722	3.175	3.629	4.082
10	4.536	4.990	5.443	5.897	6.350	6.804	7.257	7.711	8.165	8.618
20	9.072	9.525	9.979	10.433	10.886	11.340	11.793	12.247	12.701	13.154
30	13.608	14.061	14.515	14.969	15.422	15.876	16.329	16.783	17.237	17.690
40	18.144	18.597	19.051	19.504	19.958	20.412	20.865	21.319	21.772	22.226
50	22.680	23.133	23.587	24.040	24.494	24.948	25.401	25.855	26.308	26.762
60	27.216	27.669	28.123	28.576	29.030	29.484	29.937	30.391	30.844	31.298
70	31.7513	32.205	32.659	33.112	33.566	34.019	34.473	34.927	35.380	35.834
80	36.287	36.741	37.195	37.648	38.102	38.555	39.009	39.463	39.916	40.370
90	40.823	41.277	41.730	42.184	42.638	43.092	43.545	43.998	44.453	44.906
100	45.359	45.813	46.266	46.720	47.174	47.627	48.081	48.534	48.988	49.442

FEET TO METERS

ft	0	1	2	3	4	5	6	7	8	9
	m	m	m	m	m	m	m	m	m	m
-		0.305	0.610	0.914	1.219	1.524	1.829	2.134	2.438	2.743
10	3.048	3.353	3.658	3.962	4.267	4.572	4.877	5.182	5.486	5.791
20	6.096	6.401	6.706	7.010	7.315	7.620	7.925	8.230	8.534	8.839
30	9.144	9.449	9.754	10.058	10.363	10.668	10.973	11.278	11.582	11.887
40	12.192	12.497	12.802	13.106	13.411	13.716	14.021	14.326	14.630	14.935
50	15.240	15.545	15.850	16.154	16.459	16.764	17.069	17.374	17.678	17.983
60	18.288	18.593	18.898	19.202	19.507	19.812	20.117	20.422	20.726	21.031
70	21.336	21.641	21.946	22.250	22.555	22.860	23.165	23.470	23.774	24.079
80	24.384	24.689	24.994	25.298	25.603	25.908	26.213	26.518	26.822	27.127
90	27.432	27.737	28.042	28.346	28.651	28.956	29.261	29.566	29.870	30.175
100	30.480	30.785	31.090	31.394	31.699	32.004	32.309	32.614	32.918	33.223

HEAT EQUIVALENTS	
1 Btu	= 252 calories (cal)
1 Btu	= 1054.4 J
1 kcal	= 1000 cal
1 kcal	= 4.1840 Kj
1 kcal / kg	= 4.1840 kJ / kg
1 Btu / lb.	= 0.5556 kcal / kg
1 Btu / lb.	= (4.1840 kJ / kg) / (1.8 °F / °C)
	= 3.23244 kJ / kg
1 kcal / kg	= 1.8 Btu / lb.

STANDARD AIR - STANDARD CONDITIONS		
Condition	U.S. Conventional System	Metric system
Pressure	29.92" Hg = 14.696 psia	760 mm Hg 101.28 kPa
Temperature	69.8 °F	21 °C
Specific Volume	13.33 ft^3 / lb.	0.833 m^3 / kg

ENERGY EQUIVALENTS - U.S. CONVENTIONAL	
1 Btu	778 ft/lb. 252 gram-calories (g/cal) 1054 joules (J)
1 horsepower	33.000 ft/lb./min 550ft/lb./sec. 746 Watts 2545 Btu/hour 42 Btu/min. 1.014 hp (metric)
1 horsepower hour	1 hp for 1 hour 1.980.000 ft/lb. 746 Watts/hour 746 kWh 2545.6 Btu
1 Watt	3.414368 Btu/h
1 kilowatt (kW)	1000 Watts 1.34 hp
1 kilowatt hour (kWh)	1kW for 1 hour 1000 W/h

ICE MELTING EQUIVALENT TON OF REFRIGERATION
1 TON OF REFRIGERATION = 288.000 Btu / 24 hours 12.000 Btu / hour 200 Btu / min 83.3 lb. of ice / hour 3.515 kW

AREA EQUIVALENTS

1 in^2	$= 0.0065 \text{ m}^2$	
1 ft^2	$= 144 \text{ in}^2$	$= 0.093 \text{ m}^2$
1 yd^2	$= 9 \text{ ft}^2$	$= 0.836 \text{ m}^2$
1 yd^2	$= 1292 \text{ in}^2$	

VOLUME EQUIVALENTS

1 in^3	$= 0.016 \text{ L}$	
	$= 16.39 \text{ cm}^3$	
1 ft^3	$= 1728 \text{ in}^3 = 28.317 \text{ L}$	$= 0.0283 \text{ m}^3$
	$= 7.481 \text{ gal} = 28,371.00 \text{ cm}^3$	
1 yd^3	$= 27 \text{ ft}^3$	
	$= 46.656 \text{ in}^3$	
1 gal	$= 0.1337 \text{ ft}^3 = 3.79 \text{ L}$	
	$= 231 \text{ in}^3$	$= 3785 \text{ cm}^3$
1 L	$= 61.03 \text{ in}^3 = 1000 \text{ cm}^3$	
	$= 0.2642 \text{ gal}$	

LIQUID MEASURE EQUIVALENTS

Liquid measure	U.S.	Metric
1 pt	$= 16 \text{ oz}$	$= 0.473 \text{ L}$
1 qt	$= 2 \text{ pt}$	$= 0.946 \text{ L}$
1 qt	$= 32 \text{ oz}$	
1 gal	$= 4 \text{ qt}$	$= 3.785 \text{ L}$
1 gal	$= 8 \text{ pt}$	
1 gal	$= 231 \text{ in}^3$	
1 ft^3	$= 7.48 \text{ gal}$	
1 gal	$= 8.34 \text{ lb. water}$	
1.136 qt		$= 1 \text{ L}$

PRESSURE EQUIVALENTS

1 atmosphere	= 29.92" Hg	= 101.28 kPa
	= 14.696 psia	= 760 mm Hg
	= 33.94' water	= 10.33 m water
	= 2116.35 lb. / ft²	
1 psia	= 0.068 atmosphere	= 144 lb. / ft²
	= 2.036" Hg	= 70.3 cm water
	= 51.7 mm Hg	= 6.9 kPa
	= 27.7" water	= 2.307" water
1 psig	= 15.7 psia	= 108 kPa
0 psig	= 14.7 psia	= 101.3 kPa
1" Hg	= 0.0334 atmosphere	= 3.386 kPa
	= 0.491 psig	= 25.4 mm Hg
	= 1.13' water	= 0.3453 mm water
	= 13.6' water	= 70.73 lb. / ft²
1' water	= 0.0295 atmosphere	= 2.9985 kPa
	= 0.434 psig	= 22.42 mm Hg
	= 62.43 lb. / ft²	= 0.305 m water
	= 0.03 atmosphere	= 0.883" Hg
1 oz / in²	= 0.128" Hg	= 1.73" water
1 lb. / ft²	= 0.007 psig	= 0.048 kPa
	= 4.725 x 10⁻⁴ atmosphere	
	= 0.01414" Hg	= 0.359 mm Hg
1 kg / cm²	= 14.22 psig	= 10 m water
	= 2048.17 lb. / ft²	= 97.98 kPa
	= 28.96" Hg	= 0.967 atmosphere
1 m water	= 1.42 psi	= 73.55 mm Hg
	= 204.8 lb. / ft²	= 9.78 kPa
	= 2.896" Hg	= 3.28' water
	= 0.097 atmosphere	
1 mm Hg	= 0.019psig	= 0.039" Hg
	= 0.001316 atmosphere	

VELOCITY EQUIVALENTS

1 mi. / hr	= 1.47 ft / sec	= 1.61 km / hr
	= 0.45 mi. / sec	
	= 0.87 knot	
1 ft / sec	= 0.68 mi. / hr	= 1.1 km / hr
	= 60 ft / min	= 0.305 m / sec
	= 0.59 knot	
1 m / sec	= 3.28 ft / sec	= 3.6 km / hr
	= 2.24 mi. / hr	= 1.94 knot
1 km / hr	= 0.91 ft / sec	= 0.28 m / sec
	= 0.62 mi. / hr	= 0.54 knot

FLOW EQUIVALENTS

1 ft^3 / min.	= 7.481 gal / min.	= 28.317 cm^3 / min.
	= 499 gal / hr.	= 28.32 L / min.
		= 1700 L / hr.
1 ft^3 / hr.	= 0.0167 ft^3 / min.	= 0.472 L / min.
	= 0.1247 gal / min.	= 28.317 L / min.
	= 7.481 gal / hr.	= 472 cm^3 / min.
1 gal / min.	= 0.1337 ft^3 / min.	= 3.79 L / min.
	= 8.022 ft^3 / min.	= 3785 cm^3 / min.
1 L / min.	= 0.0353 ft^3 / min.	= 1000 cm^3 / min
	= 2.118 ft^3 / hr.	
	= 0.2642 gal / min.	
	= 15.852 gal / hr.	

Fahrenheit, Rankine, Celsius and Kelvin

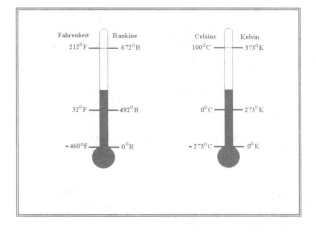

WIND CHILL TABLE

Source: National Oceanic and Atmospheric Administration

Wind Chill Index: (equivalent temperature)
Equivalent in cooling power on exposed flesh under calm conditions.

Degrees (Fahrenheit) → Wind (MPH) ↓	35	30	25	20	15	10	5	0	-5	-10	-15	-20	-25	-30	-35	-40	-45
0	35	30	25	20	15	10	5	0	-5	-10	-15	-20	-25	-30	-35	-40	-45
5	33	27	21	19	12	7	0	-5	-10	-15	-21	-26	-31	-36	-42	-47	-52
10	22	16	10	3	-3	-9	-15	-22	-27	-34	-40	-46	-52	-58	-64	-71	-77
15	16	9	2	-5	-11	-18	-25	-31	-38	-45	-51	-58	-65	-72	-78	-85	-92
20	12	4	-3	-10	-17	-24	-31	-39	-46	-53	-60	-67	-74	-81	-88	-95	-103
25	8	1	-7	-15	-22	-29	-36	-44	-51	-59	-66	-74	-81	-88	-96	-103	-110
30	6	-2	-10	-18	-25	-33	-41	-49	-56	-64	-71	-79	-86	-93	-101	-109	-116
35	4	-4	-12	-20	-27	-35	-43	-52	-58	-67	-74	-82	-89	-97	-105	-113	-120
40	3	-5	-13	-21	-29	-37	-45	-53	-60	-69	-76	-84	-92	-100	-107	-115	-123
45	2	-6	-14	-22	-30	-38	-46	-54	-62	-70	-78	-85	-93	-102	-109	-117	-125

(Wind speeds greater than 45 mph have little additional chilling effect)

How Cold is Cold? Temperature and wind both affect the heat loss from the surface of the body. The effect of these two factors is expressed as an "equivalent temperature," which approximates the still air temperature which would have the same cooling effect as the wind and temperature combination. For example, from the table above, with a temperature of 20° F and a wind of 20 mph, the effect on exposed flesh is the same as -9° F with no wind.

HELPFUL MATH NOTES

TO FIND CIRCUMFERENCE:

Circle—Multiply the diameter by 3.1416.

TO FIND AREA:

Circle—Multiply the square of the diameter by .7854

Rectangle—Multiply the length of the base by the height.

Sphere (surface)—Multiply the square of the radius by 3.1416 and multiply by 4.

Square—Square the length of one side.

Trapezoid—Add the two parallel sides, multiply by the height and divide by 2.

Triangle—Multiply the base by the height and divide by 2.

TO FIND VOLUME:

Cone—Multiply the square of the radius of the base by 3.1416, multiply by the height and divide by 3.

Cube—Cube the length of one edge.

Cylinder—Multiply the square of the radius of the base by 3.1416 and multiply by the height.

Pyramid—Multiply the area of the base by the height and divide by 3.

Rectangular Prism—Multiply the length by the width by the height.

Sphere—Multiply the cube of the radius by 3.1416, multiply by 4 and divide by 3.

DECIMAL EQUIVALENTS OF PARTS OF AN INCH

1/32 = .03125	3/8 = .3750	23/32 = .71875
1/16 = .0625	13/32 = .40625	3/4 = .75
3/32 = .09375	7/16 = .4375	25/32 = .78125
1/8 = .1250	15/32 = .46875	13/16 = .8125
5/32 = .15625	1/2 = .5	27/32 = .84375
3/16 = .1875	17/32 = .53125	7/8 = .8750
7/32 = .21875	9/16 = .5625	29/32 = .90625
1/4 = .25	19/32 = .59375	15/16 = .9375
9/32 = .28125	5/8 = .6250	31/32 = .96875
5/16 = .3125	21/32 = .65625	1 = 1.0
11/32 = .34375	11/16 = .6875	

APPENDIX

PRESSURE VS. TEMPERATURE
R-410A

Temp. (°F)	Pressure (Psig)	Temp. (°C)	Pressure (bar-gauge)
-40	11.6	-40.0	1.81
-35	14.9	-37.5	2.01
-30	18.5	-35.0	2.24
-25	22.5	-32.5	2.48
-20	26.9	-30.0	2.74
-15	31.7	-27.5	3.03
-10	36.8	-25.0	3.33
-5	42.5	-22.5	3.67
0	48.6	-20.0	4.02
5	55.2	-17.5	4.41
10	62.3	-15.0	4.82
15	70.0	-12.5	5.26
20	78.3	-10.0	5.73
25	87.3	-7.5	6.24
30	96.8	-5.0	6.78
35	107.0	-2.5	7.35
40	118.0	0	7.97
45	129.7	2.5	8.62
50	142.2	5.0	9.31
55	155.5	7.5	10.04
60	169.6	10.0	10.82
65	184.6	12.5	11.64
70	200.6	15.0	12.51
75	217.4	17.5	13.43
80	235.3	20.0	14.39
85	254.1	22.5	15.41
90	274.1	25.0	16.49
95	295.1	27.5	17.62
100	317.2	30.0	18.80
105	340.5	32.5	20.05
110	365.0	35.0	21.36
115	390.7	37.5	22.73
120	417.7	40.0	24.16
125	445.9	42.5	25.66
130	475.6	45.0	27.23
135	506.5	47.5	28.87
140	539.0	50.0	30.58
145	572.8	52.5	32.36
150	608.1	55.0	34.22
155	645.0	57.5	36.16
160	683.4	60.0	38.17
		62.5	40.27
		65.0	42.45
		67.5	44.71
		70.0	47.06